The Great Détente Disaster

THE GREAT DÉTENTE DISASTER

Oil and the Decline of American Foreign Policy

Edward Friedland / Paul Seabury

Aaron Wildavsky

Basic Books, Inc., Publishers / New York

Library of Congress Cataloging in Publication Data

Friedland, Edward.
  The great détente disaster.

  Includes bibliographical references.
    1. United States—Foreign relations—1945–
  2. Organization of Petroleum Exporting Countries.
  I. Seabury, Paul, joint author. II. Wildavsky,
  Aaron B., joint author. III. Title.
  JX1417.F74   327.73   75-3760
  ISBN 0-465-02707-5

Copyright © 1975 by Basic Books
Printed in the United States of America
DESIGNED BY VINCENT TORRE
75 76 77 78   10 9 8 7 6 5 4 3 2 1

> "It was beautiful and simple as all great swindles are."
> —O. Henry [William Sydney Porter]

Contents

ACKNOWLEDGMENTS	ix
INTRODUCTION	1
Chapter I	
THE DECLINE OF AMERICAN FOREIGN POLICY	27
Chapter II	
THINKING ABOUT AN OIL WAR	61
Chapter III	
ENERGY AND THE ASSAULT ON ECONOMIC ORDER	87
Chapter IV	
WHAT CAN BE DONE?	151
AFTERWORD	193

Acknowledgments

ED FRIEDLAND wishes to thank colleagues in the Political Science and Economics Departments at the University of California (Berkeley) whose advice he did not take often enough; their names are withheld to protect the innocent. Paul Seabury wants to thank the *New Leader*—in which an early version of Chapter II ("On Thinking about an Oil War") first appeared—for permission to reprint. Aaron Wildavsky would like to thank the Fellows and staff of the Center for Advanced Studies in the Behavioral Sciences for their help and hospitality during the time this book was written. He benefited from conversations with Ronald McKinnon, Uwe Nerlich, and George Questor who tried (but did not necessarily succeed) to set him straight.

We are all grateful to Florence Myer, who typed flawlessly under pressure; to Mappie Seabury, who maintained our equilibrium by keeping track of various versions, and improving each one; and to Midge Decter, who edited the final version with exceptional elegance.

INTRODUCTION

OIL is energy; energy is money; money is control; control is power. Oil in the wrong hands is money misspent and control corrupted; control corrupted is power abused; power abused is force misused. With oil out of control, force follows. With force out of control, so may be the world.

What does the price of oil signify? An 1100 per cent increase? All at once? A little thing, really: a mere 2 per cent increase in the cost of commodities, hardly 10 per cent of America's wealth in ten years. The enormous increase in the oil price is not a crisis but a catastrophe. It constitutes so large an amount as to cause a qualitative change in world systems. The crisis will not be short, nor limited in effect. The change is systemic. Its success inspires others to emulate it. Other primary-product cartels already are doing so.

Such effects extend from daily personal lives to patterns of relations among nations. They affect the quantity and quality of our lives, our standard of living, and life expectancy. They concern not government alone; they affect citizens. They affect not merely the economy but the polity; not America alone but the entire world; not just how we live but whether we and others will subsequently live.

The Great Détente Disaster

The systems we refer to are economic and political, international as well as domestic. The elements are connected. By "systemic" we understand that changes in some elements have effects on others; everyone's behavior is altered. The consequences are extensive—felt on a worldwide basis—and intensive—large in amplitude. Since the elements are tightly linked, change in one rapidly reverberates to the rest; when systems themselves become tightly coupled, alterations in one affect the others. These world systems now are growing unstable and unpredictable. Old linkages have been broken, old couplings divorced, without predictable new partners to take their place. Rather than ending in a smooth return to equilibrium, these shocks amplify to extents no one yet fully understands.

Our book seeks to comprehend what is happening, to predict what might happen, and to suggest preventive measures. We dislike what we foresee for ourselves as individuals, for our country, and for most of the rest of the world. Let us start where the oil price increase begins, with the international and national economies.

An immediate consequence of the oil price increase has been mass starvation in poor countries. Foreign exchange is used up purchasing oil. Poor nations now pay more in additional charges for oil than they receive in foreign aid. Oil goes into making and transporting fertilizer. The poor are caught in a double bind: they cannot purchase the fertilizer they need to grow food; the cost of transporting food from other countries has more than doubled. India, central African countries, and many others will be unable to feed many of their people.

Introduction

An oil rise also fuels inflation in rich countries. The direct costs are not the half of it. An inflationary psychology is spreading. Each economic actor seeks to protect itself by adding a mark-up to every price increase. The result is not, say, to add 10 per cent on top of the oil increase but to keep multiplying that mark-up at every step from the supplier of raw materials to the manufacturer, wholesaler, distributor, and retailer. The original 2 per cent quickly becomes 3. It then becomes part of the justification by which wages are increased and inflation keeps going. There is still a substantial difference between a single- and a double-digit rate of inflation—the difference between a rate of inflation barely in hand, and one going out of control. But this is not all.

The flow of oil generates an even larger outflow of money from importers to exporters. Importers run up huge debts. For industry this means that less money is available for investment, leading to production, that would soak up available purchasing power. For government, oil debts mean huge balance of payments deficits, which they must finance by selling assets, borrowing, or by monetary manipulations. Their temptation to run the printing presses overtime is as overwhelming as it is ultimately self-defeating. Oil adds to inflation, then, in many ways besides its direct costs. Although oil was not responsible for stoking the fires of inflation in the first place, pouring oil on inflation is about the worst thing that could be done. The problem now is not how inflation got started but how to end it, and anywhere one looks the slippery slope is covered with oil.

The Great Détente Disaster

Inflation could be fought by recycling oil dollars as investments in productive countries with capital shortages. Unfortunately, this is unlikely to happen. More than money is involved; there are problems of control, of power. Oil exporters will have at their disposal approximately $60 billion a year in liquid capital after having consumed another $40 billion themselves. This sum is small in relation to the total wealth of the oil importers but huge in proportion to the capital available for investment. Fears of foreign control are bound to grow. Capital-deficit nations may welcome oil money in theory but are likely to oppose oil investment in practice. No country wants its media of information, its steel mills, its defense industry, and who knows what else controlled by foreigners, especially foreigners who are suspected of having interests incompatible with those of the host country. The oil cartel is bad enough. And, as inflation worsens, the value of stocks plummets. Who wants to hang out a sign saying "My country for sale, one-third discount?"

Oil money might be welcomed if it brought new industry instead of buying out old. But there are formidable obstacles in the way. Domestic producers are likely to complain about competition. Why should they send money abroad, only to find it financing their commercial foes at home? Just as small business legislation in the United States includes protection against competition with existing businesses, so will legislation be sought against massive foreign competitors. The largest obstacle, however, is likely to be the fears of oil exporters about expropriation. If they are identified as full owners, they risk suffering exactly the same treat-

Introduction

ment they have meted out to the international oil companies. So they will try to keep a low profile. Their money, however, is so massive it can hardly be kept hidden.

If stationary investments become too tempting a target for interference or expropriation, moving money around is likely to become a preoccupation of oil producers. They will keep large sums on short-term deposits. They will play the currency and commodity markets seeking short-term gains. Inevitably these monetary maneuvers will prove destabilizing. Banks never will know how much they have for how long, or how much they can lend to whom, because their financial picture will keep shifting as money is moved in and out. To prevent constant chaos host nations will try to impose penalties (negative interest rates) on short-term deposits. At the same time, however, the oil exporters will not be prepared to meet the conditions imposed by investors for long-term deposits.

Like any prudent investor, OPEC (Organization of Petroleum Exporting Countries) will want safeguards for its money. It will want to invest in countries with rich and stable economies. Why hold on to currencies subject to rapid depreciation? Hence the "Third World" will be left out because they are poor; various European nations will be turned down because they are becoming unstable. To them that hath will be given, as Matthew said, and them that hath not will lose a further opportunity. Those who need most will get least.

OPEC wants what no other nations have—protection against inflation and guarantees against expropriation. OPEC wants what everyone wants, a secure future as well as an affluent present. OPEC wishes not just a dazzling decade, but a stupendous century. The trouble is not with these understandable desires but with their consequences for the other parties to the transaction.

Compensating oil exporters means condemning importers. It means that importers must accept more, in terms of higher petroleum prices, of the very phenomenon against which they are struggling. It means also that they bear all the costs while OPEC gets the benefits. Having contributed to inflation, OPEC wants to dissociate itself from the consequences of its own acts. That is a good deal if one can get it.

It would be possible, though undesirable, to provide automatic mechanisms for allowing oil prices to rise with those of other basic commodities. But how could OPEC be guaranteed against expropriation? It could get the same kind of paper guarantees from host countries that some of its members already have violated, but these guarantees would be rejected. It could try to recycle its funds through a special international agency. No matter how large a voice OPEC had in this agency, however, nothing could stop member nations from withdrawing individually or collectively if they found this to be to their advantage. The only guarantee against expropriation of foreign assets in theory is control of domestic government in fact. And sovereignty is the last thing governments and their citizens would be willing to give up.

Thus far our focus has been on what OPEC could or might do; now it is time to look at how other nations

Introduction

might be expected to react. They will be torn between the need for OPEC's resources and dislike of the conditions attached to them. They want money, to be sure, but they do not wish to give up control. They cannot exist as short-term depositories and they do not want to give long-term guarantees. As their economies worsen, they will alternately be attracted by the prospect of fresh infusions of capital and repelled by realizing that the cure of foreign control may be worse than the disease of lowered living standards. By themselves how can they alleviate their distress?

Some can try to emulate OPEC. They can, if they are fortunate, use their own commodities to form cartels and drive up prices to match what they must pay OPEC. The United States could do this with food, Jamaica with bauxite for aluminum, Chile might go with copper, and the devil take the hindmost. A few countries might do well, most would do poorly, but not all could even try. If resources were equally distributed around the world, everything might level out, but they are not. The most likely consequence of all this running hard merely to stay in the same place would be internal exhaustion and external disruption of international markets.

The rule of comparative advantage suggests that each country produce the thing at which it is most efficient and import that which would cost more to make at home. No longer. OPEC has taught that every nation must try, even at a financial disadvantage, to produce whatever it can at home so its supplies cannot be disrupted or suddenly spiral in cost. Most nations would be made poorer by being denied the advantages of international trade, but each would better be able to control its own destiny.

Now it should be apparent why we speak of systemic economic change. Domestic economies are converted, insofar as possible, into self-contained and self-sufficient units. International trade declines. After paying for oil, most nations lack the foreign exchange to buy abroad. Afraid to invest abroad for fear of expropriation, they must invest what is left at home for fear that essential supplies will be cut off. Trade partners change rapidly as commodity cartels are formed, fail, reform, and fail again. The international monetary system cannot withstand the disorderly markets in commodities and the subsequent unpredictably large flows. Flexible exchange rates alter too rapidly and a fixed rate cannot withstand endless rags-to-riches and riches-to-rags. Too much money is being pumped out, not enough flows in. Back to barter. Short-term exchange of commodities is still possible.

Disturbance abroad is matched by consternation within nations. Most are faced with a combination of unemployment, inflation, and capital shortage. Efforts to deal with each of these problems separately aggravate the others. Squeezing future investment capital out of a declining economy leaves less for current employment. Putting people to work by governmental action adds to inflation without increasing productivity. Curbing inflation by withdrawing money from the economy aggravates unemployment and discourages investment. The real meaning of increasing the price of oil eleven times: a large decrease in the standard of liv-

Introduction

ing among the oil consumers must be allocated, and for some reason domestic political processes do not seem to be able to come up with widely acceptable means to this regrettable end.

Who, in other words, will pay? That is the question. Will the costs of the oil increase be paid by the rich or the poor, young or old, black or white? Pouring oil on old wounds reinflames the scar tissue. Whatever the central cleavages in a society may be—class in Britain, race in America, language in Ceylon, caste in India— the effects of oil will be to enlarge them.

The ever-closer interdependence between foreign and domestic affairs will lead to more government intervention in the economy and society. If there is a 30-cents-a-gallon tax on the price of gas or its equivalent in import duties, or if there is rationing and/or restriction of imports, are these domestic or foreign policies? Naturally, they are both. If there is increased investment in atomic energy to assure future supplies, it may affect the health and safety of millions at home as well as constituting a counter to oil manipulations abroad.

Issues that might have been resolved reasonably, had they been confined to domestic interests, may become intractable when foreign elements are introduced. Conservationists and producers will disagree more sharply over strip-mining, oil shale, nuclear power plants, and the like. Developments that used to take decades will be speeded up to years. As the need for energy becomes more desperate, so will the conservationists who fear the ugliness of strip-mining or the thermal pollution of atomic energy or the air pollution of coal. Contradictory demands on government—produce more energy with less damage to the environment at lower cost—will increase. So will conflicts between social services and defense as the income available to both declines. As foreign news becomes less bearable, governments may try to apply more pressure at home. Yet there is no reason to believe that people will become less attached to their lifestyles or less interested in benefits from expensive spending programs. It will be difficult to reduce defense expenditures because allies and dependents will be poorer and weaker than they were. The United States may be torn between recent memories of Vietnam (no more foreign adventures!) and older recollections of the Second World War (intervene before it is too late!). The material losses, great as they are, pale before the spiritual turmoil. People cannot understand why they are suddenly being so deprived. They are the same, the country is the same; they are doing the same work; the country has not lost its productive capacity overnight. Why, then, are sacrifices being imposed on them? What, in fact, is the worth of any man's work compared to an instantaneous income transfer of $100 billion a year? Oil-importing people do not deserve to lose their income, and oil exporters (see Chapter IV) do not deserve to gain so much. But OPEC is far away and mostly mysterious. People are more likely to find scapegoats close at hand—domestic oil companies, profiteers (real or imaginary) and, of course, politicians.

If the popularity of politicians depends on how they and their policies are doing in the world, the news is bound to be bad. Demands for help at home and abroad will increase, with fewer resources to pay for

Introduction

them and less likelihood of support being offered in return. The lot of a politician is not likely to be a happy one.

In view of these circumstances—the barest extrapolation from current events—it seems hardly likely that the United States will have to worry about a too-powerful presidency, a legislative dictatorship, judicial tyranny, or any of the other scare slogans of the day. There will be enough blame for everyone. The complaint will be that American political institutions are too weak in comparison to their responsibilities, not too strong in relation to one another.

Yet the United States is in a much better position to navigate than are the smaller ships of state in Japan and Western Europe. U.S. dependence on imported oil is much less; two-thirds is produced domestically. The opportunities for further finds on its continental shelves and for developing alternate sources of energy—the country has half the world's known coal deposits and vast amounts of oil shale—are much better. Its economy is much larger and its dependence on foreign trade much smaller. The United States is better able to generate internal resources for investment and to ward off external influences. Its domestic market is large enough to allow its economy to function despite decreasing and disrupted international trade. Its agriculture and industry are so productive that demand for them is likely to continue even on a barter basis. Its traditional allies may have a few of these advantages, but none has them all or in the same degree.

Although no one can predict the precise course of events in so many different countries, no one should doubt that they will be bad. Political instability will be the norm. There will be rapid changes in government as each succeeding administration fails to meet insuperable obstacles. Elections will not be able to decide anything because new governments will be unable to do better than their predecessors. Extremist movements will arise to exploit discontent. Democracy will be in danger. The weaker among U.S. allies will ask for help but the stronger, taken with their own troubles, will be increasingly disinclined to supply it. So the search will be on for new allies. Charges of selling out to OPEC will mingle with cries of capitulation to capitalists. As the international economic system moves from trade to barter, the international political system will move from its present fragile stability to an unstable equilibrium characterized by successive seismic shocks.

International politics after the Second World War could be described fairly as a roughly equivalent bipolar system. Both the U.S.A. and U.S.S.R. expanded their global influence. The great powers, as they were properly called, reached a series of understandings that avoided armed conflict in a nuclear era. The United States did not intervene in East Berlin, Hungary, and Czechoslovakia when these countries were beset by internal revolts. The Soviet Union backed off from West Berlin and from the emplacement of missiles in Cuba.

The 1960s brought the barest beginnings of multipolarity. Western European nations began to move away from dependence on the United States through the Common Market. Eastern European nations gradually expanded their trade and cultural contacts with the

Introduction

West. The Soviet Union had been thwarted in China. The United States had been vexed in Vietnam. There were stirrings of self-consciousness among Latin American, Asian, and African governments leading to the term "Third World." How have these developments been affected by the oil crisis?

Oil has greased the way for greater consolidation in the Soviet Bloc and for disintegration of the Western alliance. The Soviet Union is an energy exporter; the rest of Eastern Europe—with the exception of Poland, which has coal aplenty—are energy importers. Inevitably the most advanced industrial countries—East Germany, Hungary, and Czechoslovakia—depend most on oil. Of necessity, therefore, the East is less able to trade with the West: it pays more for oil and it depends for supply on the U.S.S.R. The trade of the East with the West is now, in effect, routed through Moscow. Through the same set of circumstances, moreover, Russia is made richer. It can sell its arms and its oil at higher prices. Its economy grows stronger than either its Eastern allies or Western opponents.

Just the opposite occurs in the West. The Common Market can continue only as long as its strongest members, preeminently West Germany, subsidize the weaker ones. Germany still has a foreign trade surplus. But since France and Italy (its largest trading partners) are compelled to inhibit imports, and since Britain, Japan and America have also cut back, this surplus cannot survive. The temptation of each country to export its deficit by monetary means will mount. Competitive devaluations will be difficult to avoid. Cracks in the common cause between Britain and Norway on the one side, which have new sources of oil, and the other countries, which do not, are bound to widen. Fissures developing from the issue of whether it is more feasible to make separate deals with the petroleum potentates (which France favors) or to join in a united consumer front (which America prefers) cannot be easily closed.

The United States does not have oil to burn. And it is denying the necessity to undertake the kind of drastic measures that might enable it to send other forms of energy, such as coal, to the aid of its allies. Below all the rhetoric, at the bottom line, lies the stark fact that the United States cannot help; it can only seem to hinder. When the United States asks for diplomatic support in favor of Israel, or military liaison during armed conflict, it asks the European countries to risk Arab retribution without any prospect of real return. If the name of the game is oil, the U.S.A. cannot play.

NATO is a necessity that could fast become a nonentity. The North Atlantic Treaty Organization was a military umbrella under which post-World War II reconstruction took place. Until now any conceivable conflict between guns and butter has been controlled by the fact that it has been unnecessary to choose. The first aid came from the United States through the Marshall Plan and the second from high rates of economic growth. While America maintained its arms and the Soviet threat receded in the background, military expenditures went down and domestic spending (as proportions of Gross National Product) went up. But growth meant there was enough. Now that growth is gone and decline has taken its place, the issue of guns versus

Introduction

butter (we should say oil) can no longer be contained.

Energy is enervating NATO. Each member is poorer than it was. No member wants to bear the burden alone. Arms and oil cannot be bought at the same time. Forced to choose—between a remote and problematic Soviet attack and the certain threat of domestic disruption—NATO nations will choose the immediate and insistent. Yet they will find it difficult to negotiate mutual arms reductions because their situations differ. NATO soon may become a superior source of quarrels rather than a mediator of disputes. The United States, which already bears a disproportionate burden, will be asked to pick up the slack at a time when its own economy is suffering. A repeat of the dispiriting debacle of the October war—when NATO members took turns denying the United States the use of joint facilities, is only too likely because the situation remains the same: Europe needs energy which America cannot supply. NATO will not be the last ship of hope to flounder on the shores of the Persian Gulf.

Before the war of October 1973, the Middle East was the area in which the United States and the Soviet Union had not worked out an accommodation, but it was also an area that did not truly engage the vital interests of either side. The United States was joined to Israel by ties of affection, by the affinity of two democratic governments, and by the desire to maintain a dependable ally in the Middle East. The Soviet Union wished to carve out a sphere of influence which would create dependable access to the Mediterranean, deny

American hegemony, and give it a greater voice in an area bordering its own. Survival was not at stake for either. But oil has changed all that.

Oil has become the universal dissolvent. Oil has given the United States a vital national interest in Israel. For without Israel, America would have nothing with which to bargain in the Middle East. With Israel, Arab states still have something to gain from the United States, whether it be territorial concessions or limitation on Israeli advances in a new war or adjustments in the status of Jerusalem. The absence of Israel would remove any rationale for restraint.

Why should the Arab states charge any less for oil than they can get? Just because this question is an obvious one, it does not mean that it can be ignored. Wealth opens up new vistas. The more opportunities wealth brings—opportunities not only for the wealth itself but for the power it buys—the more eagerly wealth will be pursued. Within the Arab world the unifying force of enmity to Israel will be replaced by the bonds of limitless booty. Maintaining the unity of OPEC to keep the price high would become more important than anything else precisely were the main distraction—Israel—to be removed from the scene. Israel has been, for America, a sentimental favorite; oil now has made Israel an indispensable irritant without which the oyster of Arab oil price solidarity cannot be pried open.

Nor has oil had an emollient effect on the Soviet Union's interest in the Middle East. Russia's attention has been piqued by the fact that it is getting paid in today's currency for yesterday's military hardware. That attention has been deepened by the profits it makes on its

Introduction

own oil, and by the enhanced control this gives it over the economies of Eastern Europe. The picture of a prostrate West does nothing to lessen the interest. And the prospect of sharing in the $100 billion OPEC income is by itself enough to convert a modest interest into a major one. For as OPEC grows in wealth it will grow in fear, and as it grows in fear the Soviet Union will be close at hand. Indeed, should the United States withdraw from concern with the Middle East, the Soviet Union could turn tutelage into tribute and get paid for its protection.

Why should the United States wish to withdraw? The game is now being played for higher stakes, and it is by no means certain the United States will not be bluffed out. During the October war, when great power stakes were much lower, the Soviet Union threatened to send troops after its allies had attacked America's ally by surprise. Instead of insisting that it was impermissible to introduce armed forces in a place where neither power had used them before, and where direct confrontation would then be possible, the United States insisted that Israel retreat. Would not the Soviet Union's interest in intervention have increased for the next time; would it not expect a weaker response from a weaker America? And would it not be more dangerous to do the unexpected when the stakes have been raised?

Consider the condition of all concerned. Third World countries are sympathetic to the idea of getting higher returns for their resources. Though they suffer, too, from the oil price rise they see no sign of help from the United States. Only OPEC has the resources to help them. Japan and Western Europe are in a similar situation. If they felt that supporting the United States could be successful in overcoming the effects of oil, they might join in. But this is in the realm of supposition. The fact they face is that oil will be embargoed and aid will not be available. Neutrality is the best they can offer. Hence the United States, should it help to defend Israel, will face OPEC, the Arab states, the Third World, and perhaps Western Europe, as well as the Soviet Union. Today the United States convinces itself that it is not primarily a supporter of Israel but is following an even-handed policy toward the Arabs and Israel. But this illusion must evaporate. There is a fundamental contradiction between opposition to OPEC and support for its Arab members in the Middle East. America will have to decide whose side it is on, if only because the Arabs can now exert enough financial leverage, and the Israelis sufficient military force, to make it choose. The choice will not be an easy one.

Miscalculation is the menace. The United States has raised expectations it may not wish to meet under changed circumstances. The Soviet Union has the motivation to move in. Confrontation could become a catastrophe.

Surrender is not the only solution. Driven desperate, the United States, with or without (probably without) support from Western Europe, might attempt to seize oil in Kuwait or Saudi Arabia. Then it would attract support because it had oil. But the dangers are a deterrent, from the small chance of a Soviet response to the likelihood of protracted guerrilla war. A deliberate decision need not be made. The United States could seize

Introduction

upon the next war between Israel and its neighbors to reverse the results of the previous one, encouraging Israel to move into Libya and itself seizing Abu Dhabi and/or Kuwait. Saudi Arabia is not an impossibility. The likelihood of Soviet intervention would increase, but the Israelis would bear the brunt of military action. These are hardly riskless alternatives. They would be better undertaken, if at all, after the mobilization of American opinion and the support of allies. Still the risks may not seem so rash in comparison with the rapid decline of Western economies, the consequent collapse of their polities, and the decay of their societies. It may seem a pity to let two hundred years of Western civilization go by default.

Is this not scare talk? Oil is not (or need not be) the Archimedes lever of the Western world. After all, this argument continues, the next decade may be difficult but by 1985 most Western nations will have overcome their dependence on oil or obtained their own sources. They will then be poorer than they might have been, but still richer than most men have ever been and than many nations are now. So-called advanced societies cannot be worth much if they cannot take a little adversity. Who said Western standards of living were sacred or that democracy could not be defiled? Change, any change, is good for a wicked world. Let the proud be humbled and the mighty thrown down.

It is hard to tell whether the argument is that the judgment of the oil god is just or that judgment day is still a long way off. Escapism today takes the form of a "Rip Van Winkle effect." It is as if the Western world were to awake after ten years to discover that everything was the same for the surprising superabundance of energy. The destabilizing defects of the decade disappear. This is not goal attainment but wish fulfillment.

Virtue can be a vice; placed too high, virtue can lead to the subjection of the so-called uncivilized; put down too low, it can lead to subjection of the self. Excessive self-confidence is hardly the worry of the West. Loss of self-esteem is more the order of the day. If the West is not preserved, this will most likely be because its people do not believe their culture is worth preserving. Franz Fanon, not Edmund Burke, will be their prophet. Is the West so depraved that the oil of absolution is the only remedy? What is the alternative?

If oil money were flowing from the rich to the poor, evidence of a benevolent hand might be found. If the oil exporters gave evidence of charity, humanity, justice, or any other virtue, their preeminence might be preferable. But no one believes any of these things. They are, in the main, reactionary feudal regimes, or military dictatorships, or royal despotisms. Not one is governed by the consent of its people. Not one is known to deal kindly with critics inside their countries; why should they behave better to outsiders?

If what we fear is as apparent to others as it is to us, why doesn't OPEC recognize its self-interest in limiting convulsions that would ultimately engulf it as well? A disorderly world would not be desirable for Iran, Saudi Arabia, Abu Dhabi, or Kuwait, to mention a few likely losers. Perhaps they have discovered a new principle of international relations—richer is better—and wonder why others try to persuade them that poorer is

Introduction

better. Perhaps they miscalculate. Perhaps they think they can control the consequences of their present policy. They will, however, be wrong. They never had much order to begin with, anyway, and most of their people will not notice the difference. No one asked them what they thought before and no one will ask them now. They did not share in the wealth before and they can still participate in poverty now. The only difference is that if misery loves company, they will be not so alone.

One should never discount dreams, especially dreams of glory. The Shah of Iran dreams of restoring the glories of Persia. The King of Saudi Arabia foresees a Moslem renaissance. Visions of redressing ancient wrongs and modern misdeeds abound. Risks, then, may be worth running for the dreamer. It must be pleasant, after the real and imagined insults of subservience, to rise up and be importuned by finance ministers with outstretched hands and supplicant eyes, to assure these finance ministers sincerely that prices are too high only to watch their Western faces drop when a mere 30 cents-per-barrel increase, less than a penny a gallon, costs them around $4 billion a year. How sweet it must be! The taste lingers. Maybe, after so exquisite a design, the chances of being caught in one's own web are worth taking because the weaving of it is so wonderful.

Oil is working its way through the warp and woof of established relationships. West and East vie to sell arms anywhere and everywhere they can bring a return. Iran is becoming a vast arsenal. Nuclear reactors are being sent from France, with what safeguards no

The Great Détente Disaster

one knows. Money flows to this insurrectionist movement and that established regime. Revolts and repressions may be financed by certain members of OPEC without public knowledge of who is responsible. The lesson (for anyone disposed to challenge OPEC's position) will be that it is best to stay on the safe side. As the European Economic Community declines and NATO falters, fear and foreboding will increase. As governments fall and are replaced by new regimes, who can predict what they will do or where they will go? The old international order will evidently have declined, but it is doubtful whether the new disorder will be less dangerous.

The Russians may abandon detente. If detente means what Secretary of State Kissinger says it does—not cooperation but coexistence, conflict being pursued up to but not at (nuclear) sword point—then the term is merely a new synonym for the cold (instead of hot) war that has existed since shortly after the Second World War. Maintenance of that tacit understanding has all along required that neither side be tempted irresistibly to take advantage of the other. If the threat of mutual destruction is the immovable object, then the threatened loss of oil may be the irresistible force. Settling for the status quo is one thing; haggling over small advantage is another; giving up great gains is something else altogether.

Since the Second World War the nuclear powers have exchanged pieces, taken and lost pawns, always avoiding attacking the center. Whether the Soviet Union will rest content to wait and wear down its opponent in the middle game or whether it will be tempted to rush to-

Introduction

ward checkmate remains to be seen. The fact that changing the rules of the game is now a possibility shows how quickly the last quarter century of systemic stability has been subverted.

How does a stable equilibrium become unstable? A useful metaphor is the dual-control electric blanket with a difference—it can make one cold as well as hot. The system is stable so long as energy flows equally to both parties and the connections between the elements work. Adjusting the controls requires no coordination. He can become cooler and she warmer at will without affecting mutual relationships. Suppose, however, that the wires are crossed. Immediately the system becomes unstable. As he tries to get cooler, he makes her too cold, and as she tries to get warmer she makes him too hot. Left unattended, the system goes into an unstable (though symmetrical) equilibrium: the one becomes frozen and the other burns up. Energy just flows in the wrong direction. Coordination is required. Recognizing a mutual interest in accommodation, the two parties find out what is wrong, rewire the elements, and bring their system back into equilibrium.

Let us now make a different assumption: energy is uneven; power is diverted from one to the other. She is able to keep warm but he cannot keep his cool. The system has become asymmetrical. He demands restoration of mutual accommodation but she prefers to keep things as they are or, more accurately, as they are becoming. His last chance, he thinks in his weakened state, is to equalize energy by pulling the plug so they will both have an interest in restoring the one thing on which they might agree—the old equilibrium. She wonders whether increasing the energy drain might not make him too weak to shut the whole system down. Whether this story has a happy ending depends on whether the storyteller prefers the old system, the new system, or no system at all. These are the questions the oil crisis now forces us to answer.

How did the United States get into this fix? How can we get out of it? The first chapter, "The Decline of American Foreign Policy," shows that our difficulties are of longer standing than we think. The second, "Thinking About an Oil War," analyzes the crisis in political systems. "Energy and the Assault on Economic Order," the third chapter, explains what happened to oil, why it happened, and the fundamental issues that remain to be resolved. The fourth and final chapter, "What Can Be Done?" begins with a discussion of moral issues: Is it right to drive down the price of oil? Does America have the will as well as the might to act? These questions we answer affirmatively, though not without perplexity. Hence we will outline a series of alternatives from the mildest (insulation of the U.S., economic countermeasures) to the most severe (seizure of assets, countercartels, threats, force, etc.). The pros and cons of each position will be debated. We will propose a graduated series of steps from the mobilization of opinion for a drastic curtailment of oil use, to government control of oil imports, to economic and diplomatic pressures. These should be given a chance to work. If they do not, then we will discuss more forceful contingencies.

CHAPTER 1

THE DECLINE OF AMERICAN FOREIGN POLICY

> PORT OF SPAIN, Trinidad (AP)—A 26-year-old man, carrying in his pocket a note reading: "Animals should be free," slipped into a zoo cage and was killed by two lions he apparently intended to let out, police reported today.
>
> The man was identified only as Humphrey, an unemployed motor mechanic. Police said he died on the way to hospital.
>
> A spokesman said the man entered the cage unnoticed and was apparently caught by the lions before he could free them.
>
> The note he carried read: "God is love and love is life. Animals should be free."

TODAY America's foreign relations with its most important allies—Japan and Western Europe—are at their lowest ebb. On the one hand, the United States has shown it cannot help them when their standard of living is threatened by outside forces; on the other hand, these nations severally have shown they cannot help themselves. America's hemispheric relations with Canada and Mexico, are, rather than bad, indifferent. These neighbors don't care how much their oil policies hurt us; on our side, we pretend not to notice. Old friends may fade away but ancient enmities linger on. Detente with the U.S.S.R. is far from a reality. As far as China is concerned, the only American policy which it supports is our disapproval of Soviet attacks on its mainland. The United Nations seems united mostly in opposition to the United States. The record, in other words, is gloomy.

Why have the celebrated successes of the late Sixties and early Seventies suddenly turned into the Wake of the West? The answer lies not in change but in continuity.

A good place to begin analyzing the decline of American foreign policy is the India-Pakistan war over Bangladesh, because all at once it revealed weaknesses that were to show up more slowly elsewhere. Once it became apparent that India was determined to create yet another mendicant state and that the United States could not (because of lack of support at home as well as of our military requirements in Asia) take forceful action, the only proper policy would have been silence. Faced with another abominable Asian war, a dignified disapproval was the most our government should have allowed itself. Instead, it made a bad scene worse by creating gratuitous hostility between India and the United States.

An expressive focus on the feelings of top policy makers, instead of on the substance of the issues—as if *they* and *their* emotions were the central consideration—was to mark more than one episode in recent American foreign policy; Vietnam is a good example.

By 1969, when President Nixon assumed office, the situation in Vietnam was clear: to force the North Vietnamese army out of South Vietnam would have required nuclear weapons, an invasion of the north, or a doubling or quadrupling of American ground forces. None of these alternatives was morally defensible or politically feasible. It was unlikely also that the North Vietnamese could force the United States to cut off all support to the Government of South Vietnam (GSVN). The decision to take American forces out of Vietnam had been made, in effect, in the years before the election of 1968. The real questions for the administration were when and how.

Gradual withdrawal suited well both the announced policy of the administration not to abandon allies and its need to neutralize domestic opposition. As troops

The Decline of American Foreign Policy

were pulled out, the GSVN army was being strengthened; thus pressure on the North Vietnamese mounted while domestic discontent in the U.S.A. subsided. Political and economic costs to the United States decreased as they rose for the North Vietnamese.

The trouble is not that gradual withdrawal was a bad policy per se but that this administration would not follow it consistently. Instead of calm, deliberate action in all situations, no matter what the provocation, Nixon began to talk about all the (unspecified) terrible things he would do if the North Vietnamese did not behave. Personal confrontation imperceptibly replaced political relationships, as if the honor of the president, rather than the nation, were at stake.

Confusing problems of policy with those of legitimacy was, of course, a major defect of the Nixon administration from the very beginning. The problem of policy is this: to find a course of action that will meet various needs and demands. The problem of legitimacy is quite different: to maintain the right to govern. Policy called for an effective way to end the war; legitimacy demanded that the path to war's end be free of detours that would create additional doubts about government credibility.

Consider the decision to move American forces en masse into Cambodia. For the sake of argument, suppose it to be a good military choice and one with the effect of reducing pressure on American forces in nearby areas of South Vietnam: Unless the government believed that this invasion could end the war, considerations of legitimacy would require that it not undertake the action. By the time of Nixon's presidency, instead of seeing the war as a matter of military and foreign policy alone, one had to look at it primarily as involving the confidence of Americans in their political institutions. Short of starting another war, reducing that confidence was perhaps the worst thing a president could do.

With the approach of the 1972 presidential elections, and the dwindling of American troops, the stage was set for a minimally agreeable settlement. North Vietnam would no longer insist on what it could not compel—that the U.S.A. turn over the GSVN—and America also would accept what it could not change—namely, the North's presence in the South. That alone was necessary or, at least, possible. But it was enough.

Far from calling for the highest degree of diplomacy, the basic accords—withdrawal in return for prisoners—could have been negotiated by the most lowly official. The search for transitory propaganda advantage from slight differences in wording could have been handled better by lesser officials because personal prestige was not involved; each side would have been less tempted to make points in public. Americans, as every poll showed, were ready to get out of Vietnam long before their government. They were ready (moreover, they had a right) to hear that America had done all it could with its own forces and that it was wise to take its prisoners of war and go home. After a decade of American sacrifice there was no need (indeed, there was no chance) to convince anyone that a victory had been gained or that peace had been secured or even that honor was at stake.

Apart from American withdrawal, the lasting signifi-

The Decline of American Foreign Policy

cance of the Vietnam peace accords lies not in their content but in how they come about, not in what they taught the nation about war and peace but in what they taught Nixon and Kissinger about reputation and reality. It was a time when the mere show of movement brought enormous prestige without performance, as soon became evident in China.

The new relationship with the People's Republic of China has meant altering America's relations and dealings with its old ally, the other China. Big China is worth more than little Taiwan; no one doubts that, at least in the abstract. But seating the People's Republic on the Security Council imposed heavy sanctions on a U.S. ally for which there ought to have been a strong justification. Arab references to the "Taiwanization of Israel" suggest what it was the United States was giving up. But what has it gotten in return?

There is talk about somehow taking advantage of the Sino-Soviet split but no one suggests how this might be done and even less why the United States might wish to involve itself. Should the Soviet Union attack the United States with nuclear weapons, the tiny Chinese nuclear force would do us no good. Should China be the target, American commiseration ("So sorry!") could hardly help that doomed nation. Should the two Communist countries wage war across their borders, American intervention could do China little good but itself a great deal of harm. The Russians *are* paranoid about the Chinese. No doubt the U.S.S.R. *is* worried about any Chinese-American dealing; this does not automatically mean, however, that Russian unease will enhance our security.

Chou En-Lai has hinted that China has an interest in countering Soviet influence by an American presence in Asia. Preferences, of course, cost nothing. Serious desires are distinguished from preferences precisely by what the interested party is willing to give up for them. Banquets in Peking hardly seem much of a sacrifice.

Now, probably the most important country in the world to the United States in terms of increasing economic trade and political centrality is Japan. Yet its leaders were not prepared ahead for our move toward China—the first of the Nixon-shocks that still disturb American-Japanese relationships. Nothing the United States could gain from China would be worth what can be lost in Japan.

On the other hand, certainly there is a great deal to be gained from the U.S.S.R. in trying to reduce the risk of nuclear war. There is the ABM (Anti-Ballistic Missile) Treaty with the Soviet Union. What, actually, is this treaty? There appear to be so many private reservations and individual understandings that it is difficult to say precisely what the ABM Treaty contains. Why this ambiguity?

When both sides want agreement but do not quite see eye to eye, a common way out is to use vague language and/or to maintain different operational understandings of key provisions. Both processes appear to have been at work. The lack of clarity also has something to do with the fact that the ABM Treaty was just one stage (and far from the most important one) of a continuing dialogue over arms control. The provisions, therefore, take on quite different meaning if they are treated as the first of a series of accords or as the last

The Decline of American Foreign Policy

agreement the parties could reach. Let us explain.

The ABM is a defensive weapon in the most primitive sense of the term—its limited radius (say, 200 miles) means that it cannot be fired from the United States to the U.S.S.R., or vice versa. Basically it has two uses: to protect intercontinental ballistic missiles (ICBMs), or to guard cities against nuclear attack. Guarding missile sites is chiefly defensive; ABMs can protect silos against attack. They are not useful, however, for watching over empty silos after an attack has been launched. Guarding cities can be seen either way—as a means of defending people, or as a prelude to launching a surprise attack, after which the ABM would protect against a weakened retaliatory force. Essentially, the ABM Treaty allowed each side to protect its capitol (Moscow and Washington) plus one other missile site. Each side was limited also to 200 anti-missile missiles. In addition, the parties entered into a five-year offensive missile agreement that gave the U.S.S.R. 2,358 land-and-sea-based missiles compared to 1,710 for the U.S.A. The difference was justified in part by the American lead in Multiple Independently Targeted Reentry Vehicles (MIRV).

Was such an agreement worth it? Like the participants, we have to view ABM in the context of MIRV. Technological developments allowing for multiple warheads on a single missile have made arms competition more dangerous and have spurred efforts to limit their number and deployment. In essence, the agreement maintained regular missiles at existing strengths, where the Soviet Union had the preponderant advantage, but allowed competition over MIRVs—where the

United States had an advantage—to go on unabated. The result was to make it possible for the Soviet Union to catch up in MIRVs.

The estimate of the value of the ABM Treaty depends on how one predicts the course of arms control negotiations. If effective limits are placed on offensive nuclear weapons, then the ABM Treaty—for all its defects and obscurities—will seem a bold initiative whose risks will pay off in future returns. But without effective arms control, the ABM Treaty will appear as a double defeat—once for its unequal provisions, and once for the signs it gave that the United States was not fully prepared to defend its interests. That is why the Defense Department wanted more money after the treaty than before. That is why Congress passed (and the administration accepted) an amendment saying that the treaty was not to be construed as a precedent for future inequality. That is why both sides want to keep the agreement fuzzy so they can interpret it to suit future developments as they emerge. Looking at the situation from the standpoint of the bureaucratic interests involved, this generation of Soviet negotiators will have to accept a worse (i.e., less unequal) outcome than their predecessors, which may imperil their careers.

The Vladivostok Agreement, proposed by President Ford, seems to justify a negative answer to all previous questions. It does not limit offensive weapons or reduce costs or moderate the striving for superiority. In the Strategic Arms Limitation Treaties I and II (SALT), both sides agreed to limit ABMs to defense of a single site and to limit the number of nuclear weapons until

The Decline of American Foreign Policy

1977. Bombers were excluded. The preponderance of informed opinion is that the ABM limitation was meaningful but that neither side was effectively limited in the area of strategic weapons. The same, alas, is true of the Vladivostok Agreement. The number of launchers is limited to 2400, including bombers, until 1985 (1984 was apparently considered too prophetic a date). The levels have been set so high, however, that neither side is likely to have reached the set total by then. The total for MIRVs seems to have been set by adding the number in place and in the pipeline for the United States, and adding 10 per cent to reach 1320. Thus the Soviet Union, which is just beginning MIRV production, has a long way to go. Although the number of launchers is limited, the number of missiles is not; as many as will fit on a single rocket are allowed. This puts a premium on the size of launchers, and also permits competition in regard to number and accuracy of missiles.

Among the many popular misconceptions surrounding the metaphor of the arms race is the belief that the strategic arsenal of the two sides looks the same. Actually the United States has lighter and more accurate missiles while the Soviet Union has heavier rockets with warheads of greater throw-weight. Each side could believe its approach was best. Now, as the ceilings of the Vladivostok Agreement inevitably become converted into targets, the military pressures for each side to ape the other will become more intense. The United States will undoubtedly want heavier launchers and the Soviet Union more accurate ones. The ten terribly expensive Trident submarines the U.S. Navy was thinking of constructing may well become firm goals—after all, they *are* allowed.

There is no single measure of adequacy—number of launchers and missiles, accuracy, delivery, throw-weight—that captures the essence of sufficiency. In the past, each side could and did make different trade-offs. In the future, they will feel compelled to have everything that can be fitted into the outside limit of 2400 launchers. The arms competition will go on in all other ways besides numbers of launchers.

Once more, in other words, it appears that an agreement is made only for the purpose of having an agreement. Neither side is prevented from doing what it would have done anyway and both are encouraged to compete in more expensive ways. Meanwhile, back in the Middle East, the United States and the Soviet Union have been testing both each other's good will and evil intentions.

Amid charges and countercharges, claims and counterclaims, it is well to remind ourselves as literally as possible of what did happen in October of 1973. The Egyptian and Syrian attack caught Israel by surprise; 2600 Israelis died, mostly in the first days of the war. Since Israel's population of 3,000,000 is around 70 times less than that of the United States, its losses were equivalent to over 180,000 American dead—or more than three times the number in all the years of Vietnam. Resupplied with American arms, Israel regrouped its forces, pushed the Syrians out of the Golan Heights, and advanced toward Damascus while crossing the Suez Canal and cutting off the Egyptian Third

The Decline of American Foreign Policy

Army. The perilous predicament of the Egyptians led the Soviet Union to issue an urgent summons for the American Secretary of State to come to Moscow. He went. Together the U.S. and U.S.S.R. agreed to call for an in-place cease-fire, which the United States enforced on Israel (whose armies had just about run out of ammunition). The Third Army—indeed, the gateway to Cairo—was saved.

Then, in his now famous "shuttle diplomacy," Henry Kissinger arranged a truce that placed the Egyptians twenty miles on the Israeli side of the Suez Canal, which had separated the belligerents before the war. Meanwhile, Arab nations cut off the supply of oil to Japan, Western Europe, and the United States; and the major oil exporters raised the price per barrel eleven times over. Israel was isolated diplomatically as one nation after another broke formal relations or publicly supported the Arab position in return for the promised resumption of oil shipments. When it looked as if the cease-fire were not taking hold swiftly enough to save the Egyptians, the Soviet Union threatened to send in troops, or made some threat sufficiently ominous to impel the United States to issue a worldwide nuclear alert.

There is no other conclusion—America imposed a defeat on Israel. Israel did not lose on the battlefield; it lost in the diplomatic arena. Or more accurately, it lost not merely in general support but most particularly through the agency of its particular ally, the United States, who alone was capable of pressuring it into defeat.

Israel sees itself in a bind from which there is no escape: should war start again and should the Israelis lose on the battlefield, they will naturally suffer the military consequences; however, should they win on the battlefield, they feel that outside forces will intervene to deny them the fruits of victory.

But this is only Israel. The foreign policy of the United States of course cannot be administered to make the world safe for Israel. The American government also has to consider the impact of its actions on allies, adversaries, its own people, and other states in the Middle East.

Allies first. When former Prime Minister Tanaka asked for help with oil, he was told to keep a stiff upper lip, or words to that effect. Thus Japan was blackmailed into publicly changing its neutral position to one which favored the Arab cause. The Japanese economy suffered, inflation increased, and the government lost internal support. Nor was this to be the only example of one ally (Japan) forced to turn its back on another (Israel) while the United States looked on—not exactly as an innocent bystander—more as observer than mover of events.

When our West European allies, lacking the sense of participation that comes from joint consultation, realized that the United States would not be able to keep their economies going, they first threw obstacles in the way of the American effort to resupply Israel and then scrambled to make separate deals for oil, considering neither their own dignity nor America's desires. Thus no common position to resist oil price increases was possible and the twin evils of unemployment and inflation (the villainous "stagflation") achieved greater im-

The Decline of American Foreign Policy

petus. Apart from West Germany, there is hardly a single strong government in Western Europe today.

Closer to home, major oil producers of the Western hemisphere outside of the U.S.A. sought to profit from our misfortune; Venezuela raised prices and speeded up its takeover of American oil companies. And what might have been understandable for Venezuela, a nation in a poor continent, was not so easy to take from the relatively rich Canadians who combined higher export taxes with reduced allotments. This highway robbery might have been even harder to understand for those Americans who did not recall their government's recent failure to exempt Canada from a harmful import duty, the general lack of American regard for Canadian sensitivities, or the general lack of interest in our largest trading partner. The neglect of Canada has in this instance not proved to be benign.

In the U.S.A. itself, the oil boycott exacerbated an energy crisis. Because the administration had no good explanation for being caught by surprise, its credibility was further undermined. Patterns of everyday life were disrupted. Tempers rose with efforts to place or shift the blame. Prices soared. Unemployment increased. The estimated increase in fuel costs alone is around $20 billion a year—about the cost of a wartime year in Vietnam, far more than the total strategic nuclear forces, or most of revenue sharing. If anyone wanted to know where the Vietnam peace dividend went, one answer is obvious—to foreign oil suppliers. The dollar continues under pressure while the effects of two devaluations and additional agricultural exports are swallowed up by increases in the cost of imported fuel. If the conflict in the Middle East may be compared to a poker game, the U.S.A. seems lucky to have been left with its shoes on.

No imperialism, no conspiracy of the rich against the poor, no natural disaster could have wreaked as much instantaneous havoc among so many low-income countries as the geometric increase in oil prices. The new strains of miracle rice and wheat demand large helpings of fertilizer, most of which are oil-based. Thus hunger is intensified as income is reduced. The costs of transporting fertilizer to grow food, and of food to relieve famine, have about doubled. For the poor the oil price increase is not just a single event but a continuing catastrophe. More people will die because of this than because of any war fought in recent times.

Imagine what would have happened if anyone had suggested that, suddenly, the United States should spend $20 billion a year on foreign aid, or that the industrialized nations should send $50 or $60 billions to those less fortunate. The idea would have been considered ludicrous. Words like bankruptcy, irresponsibility, and revolution would have filled the air. Yet this huge transfer of resources—without any of the moral benefits linked with distribution to the poor—has been accepted with hardly a murmur.

Who benefits? The oil exporting states, mostly in the Middle East, profit most. And what will they do with this money? Part of it will go for arms, as the various deals to supply Iran, Saudi Arabia, and Libya indicate. A major consequence of the politics of oil, therefore, will be an increased supply of sophisticated weaponry, with consequences we can as yet only dimly perceive.

The Decline of American Foreign Policy

Part of the income will go for internal development, which is good; but the absorptive capacity of the nations, as they themselves point out, is extremely limited. Therefore, they must invest. Where? Only the Euro-dollar market (the U.S.A. and Western Europe plus Japan) is large enough to absorb these tens of billions. By now, paths to the doors of oil sheikhs are well worn. Indeed, their desire to dispose of this new surplus is taken as a sign of their need to cooperate with their sources of investment. This is how the transfer of ownership in American industry to oil producers came to be viewed as a benefit of American Middle Eastern diplomacy!

In compiling the list of beneficiaries, the Soviet Union and the Union of South Africa should not be forgotten. The U.S.S.R. is self-sufficient in oil for the time being but its East European allies are not; so their dependence on the Soviet Union grows. Also it holds large stocks of gold, which continues to appreciate in value as the oil-driven inflation weakens all currencies in which supply and demand operate. South Africa lacks oil but makes up for it by the increase in value of the gold it mines with apartheid black labor.

A glance at the list of winners reveals that not one is a democracy. Neither America nor any of its allies appears among them. On the contrary, most of the winners are reactionary feudal regimes whose concern for their own peoples, let alone for others, is not exactly exemplary. A few, like Libya, are avowed enemies of the U.S. or others (like the Soviet Union) are, to say the least, uncertain about which direction their future relations with the United States will take.

The Great Détente Disaster

War with the Middle East took place in the context of American efforts to maintain detente with the Soviet Union. No one can quarrel with the desirability of achieving mutual respect and avoiding mutual hostilities among the nuclear superpowers. The question, however, is whether there is in fact any detente to maintain—and whether American actions in the Middle East have in fact been well calculated to maintain it.

As for detente as such, we have for some years now been hearing a good deal of talk about Soviet restraint. It was said that the U.S.S.R. restrained itself in Vietnam; but unless restraint means only the absence of a direct attack on the United States, there is no evidence to support such a view. What we do know is that the Soviet Union supplied North Vietnam with missiles to shoot down American planes in the air and with weaponry to kill American soldiers on the ground. No definition of detente, of course, requires the Soviet Union to abandon an ally; the point is rather that the form of this support—namely, arms to aim at Americans—can hardly be construed as a positive factor vis-a-vis the United States.

Then there is the claim that the Soviet Union showed a friendly persuasion in helping to arrange the Vietnam peace accords. This claim remains conjectural because there is no information available on whether or how hard they tried. We note only that it would seem hardly necessary to use much forceful persuasion on anyone who is being asked only to return a few hundred prisoners in return for total withdrawal. But even if the question of the Soviet role in the Vietnam settlement

The Decline of American Foreign Policy

were to be held open, there is no question whatsoever about the Soviet role in the October war.

Had Israel launched a surprise attack against the Arabs—without provocation and on the most important Moslem holy day—the Soviet Union could have expected that the spirit of detente would have led the United States to restrain its client-cum-ally by demanding that Israel cease firing and go back to where it came from. So, too, had a real detente been in force, would the United States have had every reason to expect the Soviet Union to live up to it. At a minimum this would have meant that neither side would take military advantage of the other—not merely by avoiding direct confrontation, but also by not using clients-cum-allies to make war. Yet precisely the opposite took place: the aggressor was rewarded, the defender punished; and, to top it all off, the United States was attacked with the oil weapon for daring to help the defender defend itself.

When the United States itself (not merely an ally) was hopelessly mired in Vietnam, the Soviet Union did not intervene to extricate her. Instead, the Soviet Union contributed materially to make a bad situation worse. But in the Middle East Kissinger did agree to rescue not even the Soviet Union itself but a Soviet ally by calling off the Israelis when they had the Egyptians encircled. As a reward for this sacrifice, the Soviet Union acted so that the President felt compelled, just a few days later, to issue a nuclear alert. If that defines detente, then it has become a synonym for surrender—the Soviet Union gives up a hypothetical threat to send in its troops, and the United States sacrifices the actual gains of its ally.

Or, to take another glaring example, when the Soviet Union badly needed wheat, the United States agreed to supply it. When the United States needed oil, the Soviet Union encouraged the Arab states to keep denying it. There may be some definition of the term according to which this is a detente, but not if detente is to include the slightest measure of reciprocity. The Soviet Union is not to blame for getting an advantageous price on the wheat deal or even for trying to sell some back later at a much higher price; that is merely prudent self-interest. Nor indeed could the Soviet Union even be held to account for supporting the oil embargo unless the United States acknowledged in no uncertain terms that the interruption in supply and the rise in price was contrary to its interests.

Detente should not be read as requiring either party to give up vital interests, but rather to accommodate them where possible and to respect them where not. Such accommodation cannot take place, however, unless each side makes clear which interests it holds dear and is prepared to defend. Neville Chamberlain's worst sin in the 1930s was not that he believed Hitler's lies but that he deceived Hitler more than Hitler deceived him. Following a series of capitulations, Hitler had every reason to express suprise when the British finally stuck on the invasion of Poland. Any stable relationship calls for the parties to inform each other about the limits of their toleration lest the inevitable search for individual advantage inadvertently turn into mutual hostility.

The Decline of American Foreign Policy

The worst aspect of American behavior in the Middle East is that, like England in the 1930s, it clouded rather than clarified its own vital interests, and thus encouraged future Soviet miscalculations. This must have been true in an immediate sense; otherwise the United States would not have had to use the grossly disproportionate warning of last resort—the nuclear alert—to deal with a local battle in the desert between third parties. Worse still, the American failure to make clear which, if any, interests it is prepared to defend remains in force to this day. The danger is not merely that America will go on failing to protect her vital interests but that a far too belated recognition of her need to do so will set off a train of events leading the superpowers into direct confrontation.

America has allowed the intolerable to become tolerable. Rewarding aggression should not have been tolerated; letting Egypt gain twenty miles of the Sinai, and Syria, a few of the Golan heights, is the least (and most defensible) part of it. Far worse, of course, is our having allowed the Soviet Union to encourage its clients to reduce the living standards and weaken the governments of America and its allies. But the intolerable of intolerables is our capitulation to Soviet threats to send troops and/or missiles into an area where it has not been before and in which the United States had allies and interests. A more certain way of encouraging the Soviet Union to miscalculate the next time—a time when America might decide to stand firm—could hardly be imagined.

To appreciate the seriousness of American-Soviet interaction in the Middle East for future relations among nations it is helpful to consider what everyone concerned may be expected to learn from these events. As men and nations act out the lessons they have been taught by reflecting on this example of international conduct, what kind of future behavior will be encouraged or discouraged?

The poor nations of the world have learned that crime pays and that they should get in on the racket. What oil had done for the Arabs, they hope other essential goods will do for them. Whether or not the poor nations succeed in driving up the prices of what they produce, such behavior will promote economic nationalism. Every nation that can will seek self-sufficiency. The economic doctrine of comparative advantage will be replaced by a neomercantilist system in which essential commodities are produced so that others cannot deny them. Most poor countries will lose, both because they lack the most essential commodities and because the rich ones, by virtue of their wealth, will be better able to serve their own interests. The "beggar thy neighbor" league will turn into a pact for perpetuating poverty.

Arab states will learn how little they have to fear from another war. If they win, they win; if they appear to lose, a *deus ex machina* appears in the form of the threat of Soviet intervention, which is sufficient to guarantee against catastrophic or permanent loss. It must be nice to be insured against losing a war, or to play a game under the rule of heads we win, tails they lose.

The sufferings of the Palestinian people, so long without home or haven, must weigh heavily upon Arab leaders. If Palestinian lives and happiness depend on destroying Israel as a state and on occupying the land where Israelis live, there will be no permanent peace, but only continuous conflict, until one or the other is no more. Will 1973 teach them to cease war? Not if survival as a state is at stake. It can only teach the Palestinians that they were better off after the war than before. Why should they believe they wouldn't also be better off after another one?

The way in which the October war ended enabled the Arab states to settle differences among themselves at the expense of Israel, Jordan, and the United States. By recognizing the Palestine Liberation Organization (PLO) as the sole legitimate representative of the Palestinian people, and by awarding them the West Bank of Jordan, the Arab states reduced the internal pressures to deal with a multitude of guerrilla organizations. Had Egypt and Syria been forced to accept the verdict on the battlefield, going back to where they started, Arab countries would have been arguing over the advisability of accepting another defeat. Instead of fighting among themselves, they got America to fight with Israel over bargaining with PLO—whose announced aim is a Palestinian majority within Israel.

As for Israel, it cannot but be made desperate by these events. After it attacked first in 1967, following numerous provocations, Israel lost international diplomatic support. After waiting to be attacked in 1973, it suffered even greater setbacks. Israel will not learn that force alone cannot guarantee its safety, but rather that the amount of force it possessed was insufficient. Ultimately this can only mean that Israel will go for atomic weapons, for they are the only guarantee the country knows against losing all support everywhere. That such a course might be suicidal will very likely fail to prove impressive to a nation which feels the world will respect only ultimate threats.

Finally the Soviet Union has learned the secret of how to make money out of war. As the recipients of Soviet arms grow richer from the rise in oil prices, the Soviets get paid well for those arms. In turn, the Russians use this leverage to buy oil cheap and sell it dear. This indeed is a brilliant form of political alchemy—turning base behavior into gold. Because the U.S.S.R. is made to bear no blame (for a war, after all, that could not have, and cannot in the future, take place without it), they are able to reduce American income through oil price increases and at the same time to gain financial credits for importing American technology.

What does the Soviet Union have to gain from stirring up trouble in the Middle East? The answer is too obvious to allow for even a moment's speculation; it lies in a simple recognition of what has already happened: to wit, the weakening of the Western alliance. The Common Market, Japan, and the United States are poorer than they would have been, less able to bear defense burdens, angrier at one another, less cohesive internally and more divided abroad. With all this financial as well as political profit, no one need look further for plausible Soviet motivation. The only puzzle is, what is America supposed to be getting from all this?

The United States has learned to speak Orwell's language in the Middle East—defeat is victory, loss is gain, submission is supremacy. Dictating major concessions to an ally is called a diplomatic triumph.

The Decline of American Foreign Policy

Blessed are the fuel-less for they learn the joys of adversity! The American people have been taught to blame themselves for whatever goes wrong. Producers are to blame for profits, consumers for consumption, environmentalists for ecology. A 20- to 40-cents-per-barrel excess profit for oil companies is morally equated with a $9-a-barrel gain for OPEC. Everyone is blamed for doing what comes naturally at home because no one chooses to admit that there is anything unnatural abroad. Presidents cannot request sacrifices to make up for losses that never took place.

Suppose a man puts a gun in your back and cries, "Your money or your life!" and, being a prudent person, you give up your money; you lose money but keep your integrity. But if you then turned to the robber and said that this encounter had brought you closer together, that his new wealth would, by reducing his sense of inferiority, increase communication, and that, anyway, it was good for your children to learn to live without money, one of you is sick; and it is not the thief.

But what could the United States have done? Let us, once again, go back to the time when Israel was slowly surrounding Egypt's third army and the Secretary of State was called to Moscow.

If the Soviet Union needed to rescue the situation caused by the unprovoked aggression of its client, Soviet leaders should have come to Washington. The symbolic importance of these matters is not lost on them. Should America, then, have let Israel crush Egypt? No, America needed to use this opportunity to bargain for a better outcome. The minimal condition was *status quo ante*—both sides go back to where they were before. Compassion is shown for Egypt, which keeps its army and country intact, and concern for Israel, which at least is not disadvantaged by the aggression it has suffered.

The time to negotiate the oil price increase was when Israel had the Egyptian Third Army encircled. If the Soviet Union insisted on a cease-fire without conditions, the United States should have applied the same principle to demand an end to the oil boycott while negotiations proceeded. Otherwise, the United States was forced to negotiate with a gun at its head—indeed, with its economic blood dripping away. Without Soviet support, Arab oil producers would have lacked protection. If the United States insisted that its vital interests—energy, its economy, employment, its allies—were threatened, the oil producers would have had to back down or the Soviet Union would have had to back their move. But the United States asked for no concessions; rather it gave away its one advantage, the Israeli noose around Egypt, for a promise, never voiced openly, that the Soviet Union which had instigated the war would now help bring peace.

When the cease-fire did not take hold immediately, thus promising to bankrupt the Soviet policy of investing in weak clients, Russia threatened to send in its own troops. The United States had to refuse. For one thing, the presence of Soviet *and* American troops makes for confrontation if (as is only too likely) the great powers fail to agree; safety lies in both staying out. For another, neither superpower should be able to get its way by unilateral threats of force going beyond

The Decline of American Foreign Policy

areas already held. What is understandable with regard to Bulgaria is not necessarily true for Egypt. If threats of intervention became approved behavior, no one would be safe. Everyone would have to calculate whether he would be better off capitulating to the latest threat or getting in with the first one.

Furthermore, if a Soviet threat to send in armed forces is good on Monday, it is also good on any other day of the week. Acting within the context imposed by this threat would place the United States in the position of carrying out Soviet instructions; any significant deviation could reinstate the Soviet threat. Had Soviet Foreign Minister Gromyko negotiated the separation of the armies after the cease-fire, instead of American Secretary of State Kissinger, one wonders whether he could possibly have pushed the Israelis back any farther.

How could the United States have risked war with the Soviet Union over this insignificant incident? Putting the question that way suggests the U.S.A. was initiating an aggressive act against a vital interest of the U.S.S.R. But there was no such interest. Neither the territory nor the near neighbors of the Soviet Union, neither its standard of living nor that of its allies was threatened. That question is more appropriate for the Soviet Union: how could it risk a confrontation with the United States by moving troops (perhaps even nuclear weapons) into an area where they had never been? Can the U.S.S.R. have been willing to risk a nuclear war? The very idea is preposterous (though nonetheless frightening). If the Soviet Union wanted nuclear war with the United States it would not send Egypt and

The Great Détente Disaster

Syria into Israel on Yom Kippur but rather its missiles and rockets to America on Christmas. No, the Russians stood to lose the billions they invested in aggression against Israel, and they barely succeeded in bluffing their way out of a weak hand. America did not call the Soviet bluff until the next hand, when the strong cards it previously held—Israeli encirclement of Egypt—had been discarded. And then, when the nuclear alert had sounded, the American people did not understand what it was all about because their government refused (and still refuses) to tell them. What was Nixon going to say—that the proverbial pipeline had been locked after all the oil had gone?

Were the oil embargo and the subsequent price rise inevitable? Could the United States have done anything to head off these dual disasters? The past that was has a stranglehold over the future that never can be. There is a natural retrospective tendency to fit together the pieces of the historical puzzle so they make up the picture of what actually happened. To this retrospective fallacy is joined the telescopic illusion: looking backward, the actual sequence of events is foreshortened; events which occurred over weeks and days collapse almost into a single hour. Today few remember that the embargo and sudden rise in oil prices were not instantaneous; large price increases were preceded by small ones, total expropriations by partial ones, in the months and weeks before (and the days during) the war. To rid ourselves of the false telescopic view we must see the United States in the various positions it occupied over the entire relevant period of time, not

The Decline of American Foreign Policy

merely when the oil boycott was fully effective and prices had risen 1100 per cent.

More than one administration is responsible for the failure of the American government to protest the nationalizations through extortion that took place in Libya and elsewhere in the four years preceding the 1973 Middle East war. These matters should not be settled unilaterally but on the basis of consultation or through international courts. A consistent record of having taken this position amid signs that the United States was prepared to take whatever action it could against violators (from bringing suit to attaching accounts to economic measures—even if largely unsuccessful) would have helped a great deal in preparing for 1973.

While the pattern of expropriation and nationalization was clear enough, the administration may not have recognized what precedents these actions created for future embargoes and price manipulations. But there was no excuse, once war began, for not informing all concerned parties about America's determination to resist. The Soviet Union should have been told that America's vital interests were implicated and that it was prepared to defend them. The first time the Arab oil producers moved in the direction of embargo and price increases they should have been told the same thing. American armed forces could have been put on the alert; ships placed in position; airplanes making overflights, bank credits suspended, etc. Had this been done from the beginning, the Russians and Arabs might have retreated from their position, obviating the need for further tests. Instead, the embargo and price increases were accepted as normal and inevitable, and all attention was focused on how to adjust to (rather than change) them. Hence we can never know whether small actions early on could have minimized the need for large ones later.

After the fact, once the oil embargo and the price rises became parts of the existing situation, only force could have achieved an immediate change. In a different diplomatic context the task would have been easy enough—a division would have been more than sufficient to occupy Libya or any of the Persian Gulf states. Taken in conjunction with Israeli moves toward Cairo, the threat alone might have served. Saying that the days of gunboat diplomacy are over is no answer. For if America and its allies believed their vital interests were involved—domestic turmoil, international disintegration—and had acted on the premise instead of ignoring it, the response might have appeared appropriate (rather than grossly disproportionate) to the stimulus. So, too, the likelihood of Soviet intervention would have depended on its estimate of American seriousness, which in turn would have depended on whether America had convinced others about the importance of its vital interests. Since the administration had not shown this awareness, the use of force was fraught with danger. It might have been halfhearted, just enough to provoke anger and intervention, but not enough to do the job effectively. The chance of getting support in the United States for the use of force, the most important element in the calculation, would have been exceedingly low—both because the administra-

The Decline of American Foreign Policy

tion had not prepared the way, and because of the general reluctance to trust Nixon. For all these reasons, the use of force after the fact would not have been wise.

Ruling out the use (or even the threat) of force need not have ended the American reaction. There were other ways to alter the behavior of oil producers over time had there been an interest in using them. The United States might have announced a refusal to buy Arab oil until arrangements had been made for assuring supply and for negotiating a reasonable price. That would have meant continuing sacrifice at home and a crash program to develop energy resources. A tariff of, say, five or six dollars a barrel on imported oil, together with tax provisions making it desirable for American oil companies to invest at home, would have demonstrated America's unwillingness to be blackmailed and its ability to negate the effects of such bad behavior.

Instead of having to make a single difficult decision about the oil embargo, the United States will have to make endless numbers of disagreeable decisions in the future. Every time the United States has to sit down and negotiate about foreign aid, or deal with the unwillingness of Western allies to raise defense spending and the inability of poor countries to raise living standards, it will face the specter of oil price increases. Even good actions—such as the rupee settlement with India, in which billions in American-held Indian currency were prudently returned to that government—will be more than nullified by the first year or two of oil price increases.

The Cyprus affair is a good example of the impairment of American foreign policy. If America is not to be a global police force, as critics of past policies insist, then its involvement need be no greater than that of any perplexed onlooker. But, of course, the losers were not content. America inevitably was blamed for the extent of the Turkish conquest and for the humiliation of the unfortunate United Nations peacekeeping teams forced to bow to superior Turkish force. Nothing was gained by our unwanted mediation. However, had it not become clear to all in the October war that the United States could safely be disregarded, concern about American displeasure might well have led the Turks to limit their advances. By any good workable definition of "power," it became apparent that the United States had become less powerful.

What explains the decline of American foreign policy? Is it a product of necessity, the victim of past events and present occurrences? Or is it, despite appearances, a product of bad judgment? It is both. First, the time: no man, or nation, writes on a clean slate; for the most part, the world is one we never made. American foreign policy was mortgaged in Vietnam; Nixon and Kissinger did not incur this debt, but nonetheless it had to be paid out. So did the legacy of the Suez crisis. Suez was where America fought its friends—Britain, France, and Israel—and supported its enemies—the Soviet Union and Egypt. If the principle for which the British and French were fighting—no expropriation without consent or through international courts—had been confirmed by events—that is, if the United States had not intervened—the U.S.A. would have been in a stronger position to oppose the oil boycott and the hike in prices. France and Britain learned to distrust

The Decline of American Foreign Policy

America; they also learned that if the United States was not going to let Israel defend itself, next time they might as well be on the winning side. That in 1973 they succeeded only in frustrating themselves does not mean they did not try.

Just as future American foreign policy has been sacrificed to the past in Vietnam, so President Nixon's credibility was surrendered to Watergate. Trust in him had not only been mortgaged; the entire house was foreclosed. The president had become unbelievable. Where in the past legitimacy might be difficult to explain, history now offers a definition straight out of Charles Schulz's Charlie Brown: illegitimacy is when a president issues a worldwide nuclear alert and most people think he is trying to cover up some domestic misdeed.

Leadership inevitably involves an element of faith: to act in anticipation of events requires that followers believe the rationale they are given before events manifest themselves in all their seriousness. President Nixon was reduced to acting after the fact. Suppose he had to appear strong without being strong; suppose the only way to maintain credibility was never really to risk it; in an administration that cannot afford to be tested, the appearance of action must serve for its reality. Because the oil crisis could not be countered ahead of time, the country has to suffer all its aftereffects. Placing the administration in the vortex of constraints, from Watergate to Vietnam, accounts very well for its patterns of action.

The Secretary of State's maintenance of credibility for his foreign policy through his own personal credibility in the face of Watergate was an extraordinary achievement. But the price was high. It was unwise to manufacture a detente because, by doing so, they made it impossible to explain why apparent friendship turned suddenly into enmity. Foreign policy should not be made backward. The wish for detente should not be handmaid to its appearance; not its appearance justified on the grounds that to believe otherwise would be unpalatable. The danger is that the premise of the entire policy—the desirability of detente—will be destroyed in disillusionment. The ordinary, garden-variety, everyday disagreements between great powers would then be magnified by cries of betrayal, the one side because it has misled itself and the other because it has been misled into thinking the relationship would continue to be onesided.

CHAPTER 2
THINKING ABOUT AN OIL WAR

> A scorpion and a frog once met on the bank of a wide river. The scorpion said, "Let me ride on your back across this stream, as I wish to go to the opposite shore." "Why should I be so foolish?" the frog replied. "How could I be assured that in mid-stream you will not strike me with your venomous fangs?" The scorpion answered, "But why should I wish to? Were I to do that, we both would perish—you from poison, but I from drowning." "Your reasoning persuades me," said the frog, who then complied with the request. As they reached mid-river, the scorpion nevertheless stung the frog. Before both disappeared under the waves, the frog asked, "Why did you do *that?*" "You evidently are not aware," replied the scorpion severely, "that this is the Middle East."
>
> —AUTHOR UNKNOWN

THERE are now bleak incongruities in our view of world politics, and in the policy problems which arise within it. Since Vietnam, the prospect that America might use force in any other way than purely defensively is one that many people regard with dismay and horror—some because they hold the use of force to be immoral, others because they remember the civic trauma it created during the 1960s. But also we know from past experience that a paralyzed unwillingness to use force in some lesser way has often allowed situations to deteriorate to the point where infinitely greater force finally had to be used, and has all too frequently led others—severely miscalculating our basic interests and long-range necessities—to do things which otherwise they would not have done. Confusing shades of American opinion about the use of power short of war currently presents a spectacle of national indecisiveness, inviting miscalculations among friends and adversaries alike.

The same applies to many recent scholarly views about the limitations of power. There are some experts who, skirting the question of morality, argue that we are in a new age in which force no longer can succeed. "Gunboat diplomacy," with all the term conjures by way of reckless and feckless nineteenth-century adventurism, is said not to work in a world where nearly all societies have been "socially mobilized." The gaping natives who once watched with passive wonderment the approach of Western military engines are extinct. The world today is one of literate and attentive societies. Reportage of distant actions (which formerly went unnoticed) is now massive, dramatic, and instantaneous. It is not only that local populations, activated and inspired by nationalism, now have will and strength enough to resist such undertakings; but a new kind of moral disapprobation now lurks in our consciences, particularly when the sheer contrast between our evident physical might and the relative weakness of an adversary makes the contest seem unequal and, "therefore," unjust. Herein lies a contradiction: if the contest really is unequal, then why cannot the gunboat actually do its job?

Another incongruity in our views of world politics pertains to the question of the nature of the international system to which America belongs. In feistier, more self-confident times, such as the 1950s, it was a basic, widely shared tenet of American political life that this international system demanded America's vigilant attention and support. Unattended, it would be at the mercy of events and actions which might corrode and destroy it. But when Dean Rusk, testifying to a Senate committee during the Vietnam war, invoked this attitude of an earlier time ("Other nations have interests," he said, "but America has responsibilities"), many people simply shrugged. Others smirked. Now there are some among us, intent on reassuring themselves, who would argue that the international system is in any case self-regulating—that there are neat regional configurations and equilibria in operation; that local dynamics act and react on one another in such a way as to offer the benevolent possibility that America might return to a modest world role and to the pursuit of its own limited interests. Nixon's remark, three years ago, to *Time* magazine suggested this kind of comforting view of things: the shape of the emerging world system, he said, is essentially pentagonal; its effective parts are America, Russia, Europe, Japan, and China, and its centers of power essentially depolarized. In the future, the operation of the system should be conducted with prudence by all concerned. The bipolar world is, happily, a thing of the past.

Much the same holds true with respect to the idea of interdependence. Earlier, there were people who argued with eloquence the essential interrelatedness of all major political, social, and economic events; for our purposes we can call this view Globalism. Globalism too has been unmasked. A new heresy has taken its place. America's situation in the world, we hear, is one in which our relative economic self-sufficiency insulates us from traumatic developments elsewhere. By an exercise of will and the use of the right clever techniques, that self-sufficiency can be enhanced. Nixon's reaction to the winter energy crisis of 1973–1974 was a characteristic reflex action of this new idea of America's situation. Given the determination, he said, the United States could attain independence as an energy consumer by 1980. No longer would it be at the mercy of events outside its control.

The Great Détente Disaster

What we have called Globalism implies an undifferentiated American interest in, or responsibility for, occurrences in any and every part of the world. Rendered plausible, to be sure, by occasional lapses of administration rhetoric, the Globalist view is a far from correct reading of the American position even at the height of our most active interventionist foreign policy. As a matter of fact, in the 1950s and 1960s there were many, many major events in world politics in which the United States was only dimly involved, if at all. The regions of pressing American interest were confined to Europe, the West Pacific, the Eastern Mediterranean, and Southeast Asia. (The only other addition to this list might have been the Congo interlude; but here the successful object of American policy was to prevent the Congo from becoming a locus of superpower confrontation.) Nor did undifferentiated Globalism show up in America's foreign-aid programs. The major beneficiaries of U.S. economic assistance, with the exception of "neutralist" India, were countries in zones of encounter with major Communist states: Taiwan, Vietnam, Western Europe (only during the Marshall Plan), Thailand, etc. Today, however, the scarecrow of undifferentiated Globalism is taken as the embodiment of what actually was. Book after book tumbles off our presses, monotonously confirming this view of our recent history and warning against its repetition. In consequence, as might be expected, many now view any major American activism outside our own back door, no matter how it is motivated, as nothing more than a lapse into our mythical old intolerable habits.

Today, however, we are beginning to hear once more

Thinking About an Oil War

of impending threats to the international system, this time so great as to cause us to listen. The warning is that we are facing consequences which are by an order of magnitude greater than any systemic shocks the world has sustained since the 1930s.

Like many occurrences in world politics, almost no one expected the turn of events presently represented in the oil crisis. Unlike most such occurences, however, this crisis is so unique as to deprive us of any precedents to consult in dealing with it. To complicate matters further, it is fraught with extreme danger.

There can be no doubt that the major danger to world peace since 1945 has been the threat of superpower confrontation. Nor has this changed, for each situation which threatens such a confrontation carries with it the possibility of escalation to nuclear war. In the so-called tight bipolar world of yesteryear, however, the self-restraint of the superpowers created checks and limitations on their behavior, and, with significant exceptions, constraints on the behavior of their clients and their subordinate allies. In a very much loosened multipolar world, however, such constraints on clients become attenuated.

It should be pointed out especially in this connection that serious conflicts beyond the confines of direct superpower interest—which have exacted enormous tolls of life, limb and property since 1945—have not been "systemic," that is, they had no significant repercussions on the international order. However ugly and however tragic they may have been, they have not altered the general configurations of world politics. The recurrent conflicts between India and Pakistan, Moslem and Hindu, have taken millions of lives since 1947; the wars, both civil and international, among the tribes and races in Africa have gone virtually unnoticed—as was true especially in the Sudan, and more recently in Burundi. Also, the endemic economic miseries (which border on Malthus's most acute prophecies) of undeveloped countries have had minimal consequences for the general international economic order. It is a stark commentary on the nature of things that millions might die of starvation and undernourishment in Africa or the Indian subcontinent, while the general network of civilized life elsewhere has gone on, almost unaffected either in spirit or in flesh.

It is a further stark commentary on the nature of things to say that in the modern Western world there has been a naked double standard: Westerners have ignored or forgiven all misdeeds wreaked upon the Third World when these misdeeds have been both committed by and suffered by non-Westerners. The more sophisticated we are as "anthropologists"—aware of the relativity of things and practices, and of the many differing ways in which societies value human life—the more iniquities we are able to condone in "others." And because this is so, we are the more able to view such dismal events with equanimity or disregard; *we* are not involved; our vital interests are not at stake in these disasters. Perhaps this is not all bad, for were our attention really to turn upon these matters seriously, there would be no end of trouble. So we turn our backs on them, and they are non-events.

In a very important sense, as the above observations

Thinking About an Oil War

imply, the world has not been interdependent at all—in the sense, that is, that significance and value attach in our minds equally to the fortunes and troubles of all peoples. To put it most bluntly, those matters which we consciously or implicitly assume to be of vital concern to us are those inextricably linked to the material and spiritual bases of our civilization. The American failure in Vietnam sprang in large part from the fact that there was so little that could be done plausibly to make the American public believe that the fate of this small and distant country involved the fate of our culture.

It is hard to know in advance when and where our genuine passions may be brought into play; and thus rational calculations about our future reactions to events often miss the mark widely. The gap is quite wide, then, between what we think our future perception of particular incidents or crises will be, and what it in fact turns out to be. But in any case we might assume with some confidence that a series of developments which threatened the very basis of our civilization would give rise to the most momentous passions, anger, and even devotion. This latent quality of public opinion in our democracy is not notably different from what it was when George Kennan wrote in 1950:

> . . . I sometimes wonder whether in this respect a democracy is not uncomfortably similar to one of those prehistoric monsters with a body as long as this room and a brain the size of a pin: he lies there in his comfortable primeval mud and pays little attention to his environment; he is slow to wrath—in fact, you practically have to whack his tail off to make him aware that his in-

terests are being disturbed; but, once he grasps this, he lays about him with such blind determination that he not only destroys his adversary but largely wrecks his native habitat.

Why then are we so amazed at our current affliction? It is because, at one and the same time, we could never have imagined that it could come about; because we see that it forebodes an assault on the most basic, fundamental, vital interests of the advanced industrial world; and because—most unpleasant—we see that this advanced industrial world of free societies currently is bereft of the traditional authority in world affairs which it had previously exerted for centuries.

What has perhaps for so long delayed the reaction of public opinion in Europe, Japan, and the United States to the implications of the oil price rise, the energy crisis, and the monetary squeeze, is that neither here nor in the other advanced societies has it been possible to link all these phenomena in a plausible chain of causation. Inflation, for instance, cannot simply be explained by the drastic oil price rise—it was well under way before the OPEC countries acted; the energy deficit is not to be put exclusively at the door of the energy-exporting nations—in any event, it was bound to come some day, and its intensity has been due of course to the extraordinary claims of high energy-consuming industry (and as the Shah of Iran has often been pleased to point out, to the notoriously high consumer habits of Western nations).

But the movement of events on the international monetary scene inevitably will change all this. As one industrial nation after another goes bankrupt, as industries falter, as depression mounts, as astronomical wealth accumulates in OPEC countries, as the political power of Arab states grows in seemingly inordinate fashion, we move into a systemic crisis the likes of which have not been known in this generation, and which is fundamental in the sense that—barring some change—an ineluctable chain of events will force nations to act in ways in which thus far they have refused to act. When words like "the fate of civilization" begin to trip from the tongues of statesmen, something fateful is afoot.

This is no "ordinary" crisis, that is, a crisis which the major powers—by their own joint actions short of the threat of force—could "resolve." No amount of resource-pooling, no concerted activities to develop alternative sources of supply, no domestic wage-price policies can change the tempo of this chain of events. No appeals to the public for "belt-tightening" will alter the fact (by then a matter of common knowledge) that an oil cartel of historically unprecedented power threatens the vital interests of the advanced world. That this cartel is principally made up of non-Western nations, that its spokesmen will cry foul each time its actions are opposed, that it may appeal to Communist bloc countries and exert great influence over underdeveloped ones (themselves prime victims of the cartel's piracy)—all this may be assumed. Very likely any stern measures contemplated by Western nations to deal with it will expose them to the usual charge that they are nothing but imperialists commencing once more their historic depredations against the non-Western world. That the foremost victims of OPEC's price gouging also happen to be non-Western will probably be forgotten the moment the first ultimatum is issued.

A moment of historical reflection will illustrate the peculiar condition in which the Western world now finds itself: would this kind of piracy by Middle Eastern princes have been conceivable forty years ago? Even ten years ago? Obviously not. But why?

Forty years ago, the Middle East was for the most part the "property" of Europe. In a sense, it was occupied territory. Its principalities, except for Persia, were constructs of British and French diplomacy. The boundaries of these "protectorates" (now independent fiefdoms) were drawn during World War I by Europeans with such names as Sykes and Picot. The Hashemite Kingdom of Jordan actually was constructed from the bottom up by the British, who furnished an unknown Bedouin chieftain with an army, deposited him in Amman, and drew a boundary around his arid domains. Those parts of the Middle East not explicitly under European control were nevertheless Europe's dependencies. European and American oil companies jostled among themselves for concessions. For Arabs, the end of Ottoman domination had been followed by European rule. Then it would have been inconceivable to imagine an uprising of Arabs which in any way would threaten the vital interests of Europe. Quite the other way around.

But now the matter must be seen in a much more fundamental way. Europe of the pre-World War II era, divided and discordant, and beset with many fundamental problems, nevertheless was the *center* of the society of nations. No longer is this so. Moreover, the

Thinking About an Oil War

New Europe has no authority and no institutions for a collective European foreign policy. The October war made visible beyond any doubt the underlying disunity of the West. Europe, whose very survival depends upon access to Middle Eastern energy, as of now has no means of guaranteeing the flow of this lifeblood. It is at the mercy of a band of nations who have shown their ability to use energy as both a political weapon and a means of financial aggrandizement. And, those who possess oil as a political weapon also have the ability, perhaps even the will, to deny (if not to destroy) it as a deterrent threat. They can—by actions taken wholly within the confines of their borders—bring industrial Europe to a halt (if not to an end). Feudal Middle Eastern monarchs reigning over less than 1 per cent of the population of the earth, they seem determined to preside over the monetary wealth of the world. That they, along with lesser OPEC countries, have concerted their designs at meetings in Vienna is an irony which ought not go unnoticed.

So the crisis was proceeded. Would such things have been imagined *ten* years ago? Probably not. Then, or so it seemed, both the world and the regional balance of power were markedly different from what they are now. The "imperial" authority of the United States had the ability if certainly not to dictate the destinies and affairs of the Middle East and the Eastern Mediterranean then at least to contain the more dire tendencies to which this troubled region periodically gives rise. Mossadegh was "managed" when he outraged oil interests; Iran was not yet the invincible power it is taken to be today. The Sixth Fleet may have been the effective discourager of Soviet intervention in the Six-Day War of 1967. Much, then, has changed in the past decade, but one thing in particular: for all circumstances other than the worst conceivable one, American naval supremacy in the Mediterranean, especially the Eastern Mediterranean, no longer exists. In the resupply of Israel during the October 1973 war, America's "dependable" allies who furnished landing and overflight rights were—Portugal and Greece. With the billions at their disposal, the Arabs may be able to buy allies.

Such are the grim vicissitudes of the situation. We have all grown so weary of scenarios that an invitation to contemplate catastrophe is not taken seriously. But the perplexing feature of this looming catastrophe is that one of the things that adds to its seriousness is precisely the apparent lack of seriousness at any of the several stages of its development. An Arab oil producer, or any OPEC devotee, may be laboring under perhaps the most fundamental misconception of all. For, such a one might think, is not oil a commodity, to be sold or not sold? And is it such a grievance under either equity or law that the producer determine a price at which he will agree to sell? The history of past grievances might intensify his conviction, perhaps, of Western hypocrisy, and might leave him with a powerful temptation both to expose this hypocrisy and to bask in the warm sun of reverse exploitation. After all, who set the prices before? If you are able to wait long enough, history provides its sweet revenge. And now the moment for this sweet revenge is come. That it could have catastrophic consequences for everyone concerned is a thought that can no doubt be stilled in the golden glow

Thinking About an Oil War

of gathering up $1 trillion or more each decade from a compliant world. The possibility of such an agglomeration of wealth falling into so few hands and with such little effort is no less astonishing than the possibility that a bilked, bankrupt, and blundering world would accept it complacently.

The logic in this drift of events is that the Western world, innerly divided in so many ways, faces the prospect of choosing between bankruptcy and force. Thus unless OPEC reverses course in its pricing policy, unless diplomacy and common sense can bring about such a reversal, the United States—the least affected of all affected countries because of its higher degree of self-sufficiency—is faced with an acutely important problem. Would it take the collapse of only one, or would it take the collapse of more, European economies to make clear the implications of this situation?

In contrast to all the numerous postwar crises, this one is distinguished by the ubiquity of the objects which it directly and simultaneously erodes. It is not confined to a continent (Europe), to a city (Berlin), to a country (Germany), to a narrow strategic issue (the Cuban missile crisis), or to a regional contention between two or more states. It permeates and profoundly deranges the entire world, and in so doing, has the most potent political implications. It is, furthermore, unless OPEC's policies are reversed, a linear crisis.

Then, too, unlike crises which develop and subside according to whole sequences of decisions, this one owes its basic nature to only one decision—the oil price hijack. Everything else ensues from it, in the passage of time. Aside from its purely economic weight, the political question raised by this crisis is what it may portend for the future of democratic institutions in free nations as they struggle to cope with the severe domestic stresses it imposes on them. Most advanced industrial societies of the democratic world, because of the inroads of an already troublesome pre-oil-price-rise inflation, have been faced with political commotion about distributive justice within their own borders. The oil-energy-monetary crisis does not simply exacerbate this political commotion; in many countries it makes any workable resolution of the problem of distribution seem wholly out of the question.

In an expanding economy, the question is: how shall a constantly growing pie be sliced? In a static economy, it is: from whom shall be taken, and to give to whom? In a disrupted economy, in which gigantic amounts of wealth are trundled off each year to the vaults of small emirates and their allies, the question is: who shall bear the brunt of severe loss of income? Who shall pay most for the decline of the nation's economy?

In the first instance, the payment exacted will be increased unemployment, accentuated inflation, reduced social services, the growing incapacity of nations to buy essentials abroad, the end of many amenities. To be sure, arrangements have already been made for some Arab states to invest their huge earnings in control of gigantic corporate enterprises in the metropolitan countries. The spread of such arrangements could surely give rise to a startling state of affairs, in which key industrial, commercial, and financial sectors of the Western world would be controlled by foreign powers

Thinking About an Oil War

on a scale never before seen in history. The swiftness of the change of power and ownership, plus the fact that in key respects the political policies of Arab nations are not congruent with the political policies of Western democracies, pose untold difficulties. The cartelization of economic power in the hands of a few Middle Eastern states and their feudal rulers makes earlier Western plutocrats—the Krugers, Rockefellers, Sassoons, Rothschilds—fade into paltry insignificance. It has been to their past credit that Western societies have used domestic law to tame and regulate the activities of "robber barons" by subjecting their enterprises to regulations in the name of a public interest. But these new robber barons now lie outside the reach of such law. Under the guise of international law they claim immunity from control and demand their sovereign rights. Indeed they have a strong case going for them. They can easily invoke long-standing Western values of respect for private property in the name of respect for what they do with their property.

Oscar Wilde was on the right track: every system may be killed by the thing it loves. Capitalism is vulnerable to capital; market economies are subject to market manipulations. Had OPEC's armies seized Western factories, the West would have known when to defend itself and whom to fight. Had the great Airship West been held for ransom, we would have known how we became poor and whom to blame. The oil hijack might have posed problems of diplomacy or force but not of morality and will. The West would not be worried about why it was no longer wealthy, and its collective conscience would not be racked with whether it was entitled to resist these wrongs. Because OPEC moved through the market, because it collected through accountants instead of armies, it has led the West into self-doubt instead of self-assertion. OPEC's oil hijack has been equated by some with Robin Hood, though that gentleman had not been noted for making (some of) the rich richer and (most of) the poor poorer. You may not be able to fool all of the people all of the time, but apparently you can hijack most of their oil most of the time, providing you dignify it with the title of ordinary, everyday, commercial capitalism.

Heretofore, with respect to the practices of Western capitalism in the countries of investment, it can hardly be said that in general the *effects* of such investment were patently extortive and punitive. Whatever case could have been made for the idea the Western investment brought harmful social side-effects to underdeveloped countries, on the whole that investment was devoted to the economic development of these lands. Factories were constructed, roads built, ports opened. The same could scarcely be observed today about Arab investment imperialism, which entails stripping the West of its own capital—either to withhold it and place it somewhere else, or to feed it back into the same Western societies. Whatever this is, it is not productive capital. What is involved is a transfer of *power* and *control*, with no possibility that, other than oil itself, anything of significance can be added to the developed economies. Oil-producing states are essentially mislabeled. They do not produce oil; rather, they remove it from the ground. Western industry provides all the technology, the refineries, the shipping facilities, to see

Thinking About an Oil War

that the product moves properly to its ultimate destinations. There are no toiling masses sweating in subterranean mines to add the value of their harsh, demeaning labor to the product, as is true for the coal industry—another fertile if neglected source of Western energy. Could these emirates, emulating past robber barons in the Western tradition, acquire a humane streak which might allow them to alleviate the plight of their millions of current victims? The answer to this is by no means certain. But if we take the Shah of Iran's recent words at face value, they bespeak as significant an indifference to the plight of Western standards of living as to those of poorer people. We are the ones who have supposedly despoiled and plundered. We must now pay, so that others, a tiny minority of historically less fortunate persons, may reap effortlessly the benefits of gigantic wealth. Carefully, for instance, the Shah identifies his victims as giant corporations, while ignoring the plight of ordinary citizens. But the implications are nonetheless clear.

To elaborate further upon the effects of this gigantic sea-change, let us imagine that the enormous transfer of wealth and power from energy-consuming nations to energy exporters hypothetically had gone to a congregation of small nations like Switzerland. Staggering questions of equity (and the questions of the domestic economic effects of this transfer of wealth) would still remain. But Switzerland is Switzerland; the transfer at least would have been made to owners in a country characterized by prudent politics, relative domestic tranquility, a reasonably fortunate geographic location, and a benign if selfish foreign policy. Unfortunately, none of the OPEC countries can begin to claim such attributes. With respect to the Arab countries, their political policies point to a revision, by violence if necessary, of the Arab-Israel status quo in the Middle East. Persian Gulf states generally have been more restrained in their actions toward Israel than have Syria and Egypt; but nevertheless, time and again they have shown their capacity and willingness to wage war. That this capacity now has grown to an astonishing extent, and will continue to do so, is a development which must be watched with the greatest of care. There is another consideration too obvious to mention but briefly: these feudal regimes—out of pace with modern political developments, either Communist or Western—rest on a precarious base. They have no tradition of democratic civility, and are not based upon the consent of the governed. They are the remnants of a late-blooming feudalism which, elsewhere in the Middle East, has been overthrown in favor of revolutionary movements and parties. Ironically, in the recent past the survival of these feudal remnants has been due in great measure to Western aid, particularly that of the United States. (It was Anglo-American intervention in Iran, in 1953, which succeeded in ousting Mossadegh and restoring the then-popular Shah to power.) But in protracted contest with their previous benefactors, now their victims, this traditional relationship will cease. What would be the effect within their own regimes of such a protracted contest? Would they stand? One would assume that the leadership of these states, aware of the possibility of domestic insurgency, might in plain self-interest behave more modestly. In their own

Thinking About an Oil War

eyes, perhaps, they think they are already behaving so. But they should know that they are not. Their proximity to the Soviet Union, combined with their fragile domestic condition, should warn them of the need for prudence. It should also be a reminder to them that their fate is closely bound to the fate of those whom they now imprudently victimize. But, like the fabled scorpion, they may enjoy the sting enough to risk sharing the fate of the stung.

It is with respect to the possible future use of Western force by desperate oil-consuming nations that a range of questions arise, all tied together, which should make us realize the character of the issues at stake. For Europe, the United States, and Japan, contemplating the use of force—whether it should be used at all, or under what circumstances it might be used—now urgently calls for the most realistic and careful discussion. But it should be talked about now, before we go much further down the road. Otherwise, we shall perennially protest against the use of force until, in a final convulsive moment, a moment that we cannot foresee (and for which friend and foe alike are unprepared), force follows anyway, blindly, from frustration.

It goes without saying that nearly all Western-style political communities today recoil from the idea of using force in their international relations. The way in which they have respectively been led to this pacific view of things has differed greatly from community to community. Europe, long an American strategic dependence, has quietly slipped the reins of its overseas political responsibilities in the past quarter century, while at the same time growing to unparalleled heights of economic prosperity. The ensuing civic culture, at which many have marvelled as a benign portent or at least example of the way in which all societies should move, has been contingent on developments outside its control. Safe to say, their essential dependence on a benign international environment has passed unnoticed by most Europeans, as it did for the United States during most of the nineteenth century, when we avoided "entangling foreign alliances" by dint of British protection. Had it been otherwise, Europeans might have troubled earlier to take measures to safeguard themselves. To be sure, in NATO some of them have become accustomed to forms of cooperation against the potential power of the Soviet Union. But not in other directions. The fact that this quintessential civilian society could be a possible prey to forces other than the now-traditional threat from the East has meant that European thinking is not accustomed to viewing its security, tranquility, and prosperity as matters to be *otherwise* vigilantly defended.

In America, the recent experience is considerably different, but the results remarkably similar. The war in Vietnam has already been mentioned. However divided they are on other matters, both the American Left and Right have heaved sighs of relief that we are out of that war. The Nixon Doctrine was taken to signify—in Nixon's own words—that an era of "negotiation" had replaced an era of "confrontation." Recently it has been hard to imagine the kinds of circumstances in which any American President would contemplate the use of force abroad, reminded as he undoubtedly would be of

Thinking About an Oil War

the likely domestic unheavals that would attend it. Congress has abetted this disposition to moderation by weaving constraints around the presidency. The military itself, mindful of its public ordeal during the war, probably for its own self-interest fears such tasks more than it fears the dangers attendant on avoiding them. It wants not merely an abundance of arms but a superabundance of supply so it cannot suffer. A guarantee of absolute victory is nowadays the same as a promise of defeat, because the price of waging war becomes prohibitive. Haunted by the immediate past, tormented by the ravages of Watergate, no administration—even the one we happen to have now—would willingly choose a course of action entailing the use of force except in the most dire and desperate circumstances.

But prudent policy should seek to prevent precisely such dire and desperate circumstances as these. One way to avoid them would be to appear credibly committed to a careful protection of our vital interests. It ought to be known to the world, for example, that we would regard it intolerable if as a consequence of the monetary-foreign-exchange crisis one or more of America's most necessary and valued allies were to suffer political collapse. It ought to be known to the world that the delicate network of monetary and commercial relations linking the advanced Western societies to one another is a non-negotiable vital interest. And it ought to be made clear that any massive threat to this interest from whatever source, if it arises from deliberate policy, would not be tolerated.

We face an extremely unpleasant prospect. Tamed statesmen, well aware of the ways in which their more determined and forthright predecessors have been flagellated and driven from office, and wishing to avoid a similar fate, may choose to let developments deteriorate to a point where it would be *politically* feasible for them to act in necessary ways. Gloomy sequences of events thus would shape the opinion necessary to cope with catastrophe. The only question would be whether the deterioration had passed a point of no return.

In the American instance, the growing myth of potential self-sufficiency is what particularly clouds the public vision. For this myth makes it possible to imagine a less than catastrophic scenario—one in which, in fact, the American economy might ultimately profit, as if inheriting a bankrupt economy is better than running a profitable firm. So long as this view of things commands some credence, a dangerously narrow sense of our situation suggests itself to people: that America—precisely because of its relative self-sufficiency—can better withstand the shocks of worldwide economic crisis; further, that major political and social dislocations in Japan and Europe would have no parallels here. But the idea that we are protectively insulated, however well it might be argued in the abstract, leaves out of account the incalculable psychological effects on us which would radiate from both Asia and Europe as the crisis widened. The prospect, in both Asia and Europe, of the collapse of Western democracies could not help but have enormous effects within the United States. The political and ideological polarization likely to attend such dislocations inevitably would be mirrored in American politics. In all these circumstances, moreover, it should be remembered that the closed systems of Communist societies will be better able to ride roughshod over such difficulties as may attend their attempts to control their economies. Governments with a record of massive repression in the past will retain their authority more successfully in times of economic disaster than will the democracies. Whatever may be Soviet problems and responses in the current situation, it is clear that the inner disruption of Western Europe would decisively tilt the balance of power eastward, possibly irreparably.

Within governments, many of these considerations are well known. Yet governments are often loath to utter what they know; to express such thoughts officially would be to risk more calculated anger from the oil-producing states than already exists. Doubtless, quiet diplomacy already is at work to impress upon adversaries the long-term cataclysmic consequences that would ensue from continuing their present policy. In all political affairs, it is important to distinguish carefully what one would wish to do from what circumstances might compel one to do. In dealings with the Middle Eastern OPEC countries, this distinction must be made particularly clear. Their own viability, after all, depends upon a viable international system—indeed, the very system whose foundations are now being shaken by their policies. This is true quite irrespective of the policies of Western nations, present or future—unless, that is, they recognize the wisdom of making major changes in their own.

Yet it is equally important that publics in the Western democracies ponder carefully the various unpleasant alternatives which now lie ahead. The metaphor of

The Great Détente Disaster

Kennan's dinosaur need not apply inevitably to people who prayerfully prepare themselves for future adversities. But it should be said that the famous posture of the ostrich invites a kick in the rear. When their heads come out of the sand, their protection is only to be found in their ability to run at great speed from their predators. This talent is hardly appropriate to present circumstances. Why this is so will become apparent as we observe what the various players are doing with the cards they hold when the international oil game is being played.

CHAPTER 3
ENERGY AND THE ASSAULT ON ECONOMIC ORDER

RECENT EVENTS of profound significance in the world oil market have gone largely unrecognized. Everyone is of course aware of spectacular increases in the price of crude oil, along with the political and economic disarray engendered by the October war and the oil embargo. But the potentially catastrophic consequences of these mounting prices seem to be beyond our capacity or willingness to comprehend. Oil-consuming nations, and all those who try to interpret the meaning of oil-market behavior, evidently have been so transfixed by recent events that they are rendered blind to the worldwide metamorphosis of which they are a part. Even when the reality has been harsh enough to force itself upon our consciousness, the impulse to treat discrete immediate problems has obscured the underlying cause of this reality and made it hard to prescribe appropriate therapy. If the incapacity to see what is happening can be explained as one symptom of our distress, it is also one of its principal causes.

A brief description of the current economic situation will set the scene. A natural starting point (and focal point of recent concern) is the price of oil. Few economic phenomena can rival the complexity of prices in the world oil market, which depend not only on the physical and chemical properties and the location of oil, but also on whether the transaction involves crude oil or refined products such as gasoline and distillate fuel. Prices vary among the domestic markets in different consumer nations, and vary also as a function of differences in patterns of prevailing wholesale and retail trade.

Even as we narrow our glance to one type of crude oil we discover that changing institutional realities impinge upon the theoretically simple idea of "price." Since World War II, for instance, the price of crude oil has been set in terms of a bewildering array of notions involving multiple tiers, base points, and—since the early 1950s—posted prices.[1] These "posted prices" have been fictions designed to facilitate administrative action and to conform with U.S. tax laws. The actual price paid by purchasers—and therefore the actual "take" received by exporting countries—depended upon such factors as the type of crude, transportation and insurance expenses, and (most important) the ability to bargain. Precise details of petroleum price transactions have been carefully guarded by those involved in order to preserve their relations with other producers or consumers and their governments. For our purposes, however, precision is not required. The enormity of recent changes in price has been such that even rough approximations to the "actual" price serve to illustrate the economic forces which have been unleashed.

As of January 1975 the "take" per barrel of crude oil received by the producing countries of OPEC is ap-

Energy and the Assault on Economic Order

proximately $10, or more than eleven times that received in 1970.[2] (Common reference to the fourfold increase is to the posted price, not to the real one.) Dollar revenues received by OPEC increased from about $3.4 billion in 1970 to about $110 billion in 1974.[3] Thus during the past four years, OPEC nations have been able to increase their revenues by a factor of more than thirty.

What impact has this had upon the major multinational oil companies and independents who export and market this oil to the consuming countries? Increased crude prices have raised the refiner's cost of raw material from about 2 cents per gallon to about 25 cents per gallon excluding transportation costs. However, prices at which refined products are sold have doubled and—for some, such as residual fuel oil—tripled; this more than compensates for the rise in the price of crude.[4] The net result has been an unprecedented increase in oil company profits. To conclude from this, however, that OPEC's action has well served the oil companies would be premature. One consequence of recent events in the oil market, particularly the response of consumer nations to the 1973 embargo and to the escalation of prices, has been to suggest that oil company investments can be expropriated with impunity. The recent profit increases represent only a pittance in compensation for the likely loss of that abundant future which the companies would have enjoyed if only their traditional role vis-à-vis OPEC nations had been preserved.

Because of the transient impact of the increased value of existing oil in storage and the inaccessibility of data on oil company operating expenses, it is hard to determine exactly how the relative profit shares of the oil companies and exporting countries have changed. In very rough terms, the portion of total profits realized by the companies from oil sales at all stages of distribution has decreased from about 25 per cent to less than 5 per cent, while the share received by OPEC governments has risen from 75 to 95 per cent.[5] But this is not the point at which we ought to examine the merit of competing claims for profit or justice. Our attention will be directed there in due course.

As far as price itself is concerned, the significance of the startling increase does not depend upon the motives or justifications of those responsible for it. Although allocation of blame may play a part in determining our recommendations for future action, we must begin with an objective review of new facts created by the price rise.

Oil is, of course, the dominant source of energy in America and worldwide. It is consumed both directly and indirectly as a source of electric power; it provides aproximately one-half of the total energy for the U.S.[6] In 1972, almost half the energy used in residential and commercial structures for space heating, water heating, air conditioning, etc., came from oil. Industrial processes, the largest domestic energy-consuming sector, derived about one-quarter of their total energy from oil. And transportation, which depends almost exclusively on oil, accounted for one-quarter of total energy used and about 55 per cent of total American oil consumption. Because of their pervasive contribution to the cost of producing, distributing, and using virtually every item of commerce, increased prices for petroleum have fueled inflation. But these increases are far more

Energy and the Assault on Economic Order

than just one more factor contributing to an already troublesome inflation.

Since oil is the largest item in international trade, the price change has had profound financial consequences for the national economies of producing and consuming states.[7] Basically, the problems created are simple enough to perceive. How can consumers raise the money to pay? And, what can exporting nations do with their burgeoning wealth? It is far from simple, though, to understand the dynamic interaction between oil price rises, monetary fluctuations, changes in world trade, and the mounting specters of a worldwide depression and starvation in the less developed countries.

To recognize the absolute magnitude of the sums involved is imperative. Table 1 indicates approximately how much money selected importing nations must provide to purchase oil from abroad. In 1974 oil imports cost the U.S. approximately $24 billion, and more than twice that for the nations of Western Europe.[8] The total oil import bill for consuming nations amounted to roughly $120 billion.[9] As Table 1 shows, the impact of recent increases in crude oil prices has resulted in a multiplying of the fraction of their total import expenditures that nations must pay for oil. Similarly, relations among the size of oil-import spending, trade balances, and reserves is clearly menacing to consuming nations. Moreover, even if present import levels can be maintained, by 1980 the cumulative bill may be in the vicinity of a trillion dollars.

The rise in oil prices is largely responsible for the major world economic events of the last four years. A

TABLE 1
Oil Import Data: Selected Areas

	YEAR	U.S.	CANADA	UNITED KINGDOM	GERMANY	FRANCE	ITALY	JAPAN	PAKISTAN	SRI LANKA	INDIA	GHANA	KENYA	BRAZIL	CHILE
Value of Oil Imports in Billions of Dollars [1]	1974	24.0	4.0	9.5	12.0	10.7	9.1	18.0	.260	.150	1.350	.070	.115	1.425	.362
	1973	9.3	1.3	3.8	5.2	3.9	3.4	6.6	.085	.050	.415	.025	.040	.540	.147
Total Imports in Billions of Dollars [2]	1974	110.7	35.6	57.9	71.9	56.2	37.1	67.1	2.0	.342	3.9	N.A.	.747	13.8	1.480
	1973	73.2	24.9	38.8	54.6	37.7	27.8	38.3	.981	.421	3.0	.450	.615	6.8	N.A.
Total Exports in Billions of Dollars [3]	1974	101.1	35.2	40.4	91.2	48.1	25.0	55.0	1.586	.426	6.4	N.A.	.554	6.6	2.208
	1973	71.3	26.3	30.5	67.5	36.7	22.2	37.0	.961	.388	5.1	N.A.	.461	6.2	1.231
Balance of Trade in Billions of Dollars	1974	−9.6	−.4	−17.5	19.3	−8.1	−12.1	−12.1	−.414	.084	2.5	N.A.	−.193	−7.2	.728
	1973	−1.9	−1.4	−8.3	12.9	−1.0	−5.6	−1.3	−.020	−.033	2.1	N.A.	−.154	−.6	N.A.

	Year														
Financial Liquidity in Billions of Dollars[4]	1974	15.3	5.9	6.8	33.1	8.4	5.4	12.9	.286	N.A.	1.439	.173	N.A.	6.2	N.A.
	1973	14.4	5.8	6.5	33.1	8.5	6.4	12.2	.479	N.A.	1.142	.189	N.A.	6.4	N.A.
Oil Imports as percentage of Total Imports[5]	1974	22.5	11.2	16.4	16.6	19.0	24.5	26.8	15.7	26.9	31.5	11.1	11.3	16.2	N.A.
	1973	12.7	5.2	9.79	9.5	10.3	12.2	17.2	5.8	9.3	10.3	4.4	4.7	7.6	N.A.
GNP in Billions of Dollars[6]	1971	1152.0	92.4	146.9	235.5	176.7	108.2	256.4	N.A.	2.1	57.4	2.1	1.725	41.2	7.540
Total Oil Import Cost 1974-80 in Billions of Dollars[7]	1971	154.0	N.A.[8]	66.5	84.0	74.9	63.7	126.0	1.8	1.1	9.5	.5	.8	10.0	2.5
Oil as percentage of Primary Energy	1971	45.4	53.4	45.6	54.0	66.4	79.7	75.2							

[1] From Walter J. Levy, U.S. Senate Committee on Interior and Insular Affairs, *Implications of Recent OPEC Oil Price Increase*, 1974. Data for 1974 are Levy's projections; U.S. data corrected to show actual oil import totals.
[2] Levy report and IMF statistics. Data for 1974 are second quarter data expressed as annual rate; first quarter data used for Italy, Sri Lanka, India, Kenya, and Chile.
[3] IMF statistics. Data for 1974 are second quarter data expressed as annual rate; first quarter data used for Italy, Kenya, and Chile.
[4] Financial liquidity source, IMF statistics. Total reserves: 1974 data as of August 31, 1974, 1973 and 1972 year-end positions, India as of April 1974, Ghana as of June 1974, Brazil as of July 1974.
[5] IMF statistics and Walter F. Levy, *op. cit.*
[6] AID data GNP 1971 billions of dollars.
[7] Future consumption projected at present rates.
[8] No total is shown since Canada is likely to be a net exporter.

series of currency devaluations, including two downward adjustments of the American dollar, were instituted by the industrialized nations in attempts to bolster their own balance of payment situations. This objective was, of course, a key factor in the rationale for imposing price controls in the U.S. From the second quarter of 1972 to the first quarter of 1974 international trade grew from an annual rate of $276 billion to one of $516 billion, as production for the export market soared in most industrial nations. At the early stage of this expansion, raw material prices increased rapidly as suppliers faced a favorable market situation. By the beginning of 1975 it became clear that diminished consumption (caused by the withdrawal of purchasing power embodied in payments to oil-producing nations) had exacerbated depressed economies in many countries. Under these circumstances we might expect a decrease in the price of some raw materials. The price of copper has followed such a path in response to sagging demand. However, such a price decrease need not follow decreasing demand. Whether or not prices fall depends upon the institutional arrangements in each market. Sellers may react by trying to coordinate their actions—curtailing output with the hope of stimulating an overall increase in receipts. In this situation the economic system becomes clearly destabilized; rising prices and decreasing real output reinforce each other. This prospect is particularly vexing because we may face a situation in which producers of raw materials strive to emulate the actions of OPEC. That is—believing OPEC has shown that the path to economic success lies in limiting production and creating a marketing cartel—their actions will promote depressions. Additionally, industrial nations may respond to this financial pressure by adopting "beggar thy neighbor" policies which involve dumping manufactured goods at ever lower prices. Of course, the fact that there are other causes of inflation and depression does not in any way minimize the overriding significance of oil. On the contrary, oil becomes even more important because it makes an already bad situation worse.

Each nation, whatever its role in the world economic system, finds itself threatened by the twin external threats of coordinated action on the part of both suppliers and customers, and threatened yet again by heightened competition with its similarly tormented competitors. Individually, in order to afford imports, each nation—by attracting investment from abroad—must generate funds, borrowing or "earning" them as positive trade balances. It is quite conceivable that many poorer nations will find it almost impossible to afford such imports at all. As the economies of the industrialized nations contract, poorer nations will discover that those nations are less willing to invest and lend money, and less able to purchase the goods exported by the poor. Agriculture, moreover, is energy-intensive. The price of food and its very availability in the poorer nations will depend upon the price of oil as reflected in fertilizer, irrigation, and transportation costs, whether food is imported or raised domestically. Perhaps the most macabre aspect of all this is the prospect that many will starve to death while the industrial nations and the newly rich members of OPEC engage in civilized debate as to who is obliged to save them.

As for the members of OPEC, the increase in oil price has naturally vastly improved their financial position. In 1974, for example, OPEC nations as a group realized a trade surplus of more than $60 billion—an amount almost equal to the total financial reserves of Western Europe.[10] By 1980 they will have a surplus of around $300 billion. At the present rate of accumulation, surplus OPEC funds will be sufficient to buy all central bank gold holdings in less than four years, a majority interest in every firm listed on all the world's stock exchanges by 1979, and complete control of all these firms in less than a decade.[11] As Table 2 shows, revenues derived from oil exporting were multiplied by a factor of five in the year 1974 alone. In per capita terms, Abu Dhabi's revenue is $48,000 compared with about $3,000 for Saudi Arabia and $150 for Indonesia.

The significance of this enormous transfer of wealth to OPEC members depends, of course, on the uses to which such wealth is devoted. Major differences exist in the capacity of producing states to assimilate funds in the form of domestic investments. Furthermore, even where such opportunities are believed to exist, political considerations, such as the desire to maintain traditional societies, will prevent their undertaking. Unfortunately, current data on the disposition of oil revenues is not available. But the very rough estimating factors employed by Levy allow for intelligent guesses.[12] Thus, in Table 2 we show a country-by-country split for 1974 indicating the portion of the trade surplus used domestically and the remainder available for foreign investment. Levy's data indicate that

TABLE 2
Oil Export Data: Major Exporters

	SAUDI ARABIA	IRAN	KUWAIT	IRAQ	ABU DHABI	LIBYA	NIGERIA	ALGERIA	VENEZUELA	INDONESIA
Value of Exports in Billions of Dollars [1]										
1974	19.4	14.9	7.9	5.9	4.8	8.0	7.0	3.7	10.0	2.2
1973	4.9	3.9	2.1	1.5	1.0	2.2	2.0	1.1	2.8	.8
1972	3.0	2.4	1.6	.8	.5	1.7	1.2	.7	1.9	.5
Population in Millions										
1974	7.9	32.4	1.1	10.7	.1	2.2	60.8	15.9	11.5	126.4
1974	9.3	6.3	.9	3.1	N.A.	3.4	3.1	2.0	4.7	1.7
1973	3.9	1.2	.5	1.6	N.A.	2.1	.6	1.1	2.4	.8
1972	2.5	1.0	.4	.8	N.A.	2.9	.4	.5	1.7	.6
Financial Liquidity in Billions of Dollars [2]										
Total Oil Exports 1974–1980 in Billions of Dollars [3]	135.8	104.3	55.5	41.3	33.6	56.0	49.0	25.9	70.0	15.4
Potential Internal Use of Revenue 1974 [4]	6.2	8.4	3.4	2.1	1.2	3.2	7.0	3.7	6.6	2.2
Resulting Surplus Funds in Billions of Dollars [5]	13.2	6.5	4.5	3.8	3.6	4.8	0.0	0.0	3.4	0.0

[1] From Walter J. Levy, U.S. Senate Committee on Interior and Insular Affairs, Implications of Recent OPEC Oil Price Increase, 1974.
[2] IMF statistics. Data for 1973–1972 are year-end holdings and data for 1974 as of August 31, 1974, Libya July 1974, Nigeria July 1974.
[3] Levy, op. cit.
[4] Potential Internal Use from Levy, op. cit., based on Levy's assumptions that (1) Nigeria, Algeria and Indonesia will employ all their oil revenues internally, (2) Iran, Iraq and Venezuela will use 3.5 times 1972 oil revenues internally and that other Middle Eastern exporters will use twice their 1972 oil revenues internally.
[5] Oil revenues minus potential internal use.

more than $40 billion, or somewhat less than half the revenues received for exported oil, were sent abroad.

There are a variety of political considerations which govern how each OPEC nation will spend its mushrooming oil revenue. The way in which such expenditures as industrial development, weaponry, social services, personal luxuries for the rulers, foreign aid, and trade credits are allocated will shape not only the domestic results within each nation but also their impact upon the entire world economic system.

The rise in oil prices has brought almost all countries to a characteristically distasteful recognition of just how much they depend upon each other's economic actions. OPEC nations, particularly those with large oil revenues and a disinclination toward rapid industrialization, must face the fact that world monetary stability depends upon their willingness to extend credit to nations whose prospects of repayment reside primarily in the ability of their economies to recover and expand. If OPEC countries want to remain aloof from such a difficult and previously unfamiliar role in the international financial community, they can do so by hoarding their newly acquired treasure in whatever form they trust. If they buy industry abroad, they risk expropriation; if they keep their money in short-term accounts, they risk losing it in the economic collapse of their debtors; and if they leave oil in the ground, they may bring about retaliatory action from desperate consumers. Under present circumstances, however, OPEC members cannot help but observe the resemblance of industrial nations to the goose that laid golden eggs. While one may shrug off the claim that some moral law

Energy and the Assault on Economic Order

calls for supporting the goose, clearly it would be irrational to exterminate the creature.

On a slightly more modest scale, the investment behavior of OPEC nations may induce systemic instability by the selective shifting of short-term deposits, by speculating in commodities, or by speculating against the currency of particular nations—either in a search for profit or, more likely, as part of a program to force governmental compliance with OPEC political objectives. Sugar and soybeans, pounds and yen, may undergo endless price gyrations. Once the power of liquid capital is unleashed, there may be no way of containing the crests of the ensuing flood.

If the predominant response of individual nations to inflation and depression is "every man for himself"—that is, engaging in competitive currency devaluations, import restrictions, and the dumping of exports—systemic instability may follow in the form of a chain reaction of destructive international competition that feeds upon itself. Each effort to make others bear the costs will make it all more expensive for everyone. Furthermore, as policy makers perceive the domestic political consequences of depression, of national vulnerability to competitive trade warfare, and the prospect of new, output-limiting, raw-material cartels, they may try to insulate their national economies entirely.

The quest for economic autonomy as a defense may generate mercantilist policies. All nations will try to produce what they need within their own borders. A desire to secure raw-material supplies, for example—even if deflected from its traditional historical, colonialist expression—could result in efforts to overcome foreign competition by the imposition of laws against exporting raw materials, foodstuffs, and technological data. Indeed, such policies may have political appeal to consumer groups and labor organizations trying to preserve the present economic order. Despite the fact that logical arguments based upon the threat of retaliation can easily be raised against the protectionist view, the *political* salience of such arguments will not be very great. For in the face of economic reversal, anti-free-trade sentiment will not be difficult to marshal. When the present world economic system is conceived to be working against U.S. interests, we can expect a wave of public demand that we pick up our marbles and go home. As Americans reflect upon this nation's postwar humanitarian endeavors, and on the fact that some marbles (in the form of capital investments abroad) cannot be taken home, their resentment and frustration, whether justified or not, may become the central fact of domestic political life.

It calls for no great leap of imagination to see that a potential wrenching of the world's economic and political order would accompany a 180° turn in American thinking about its global ties. Part of the appeal which theories of free trade have exercised upon us after all hinges on the internationalist political outlook that they both foster and serve. However, given the present economic reality, no one can fail to see that the notion of international trade as a "good thing"—that somehow or other international trade must benefit not only trading partners, but all participants in an integrated global economy—is a vain one. Certainly the view of

Energy and the Assault on Economic Order

extensive international trade as necessarily beneficial to the U.S. will come to be seen as fallacious.

At each point in time the price of oil signifies only a temporary resultant of interactive economic forces. Like the score of a ball game, it has little value for prediction and for the analysis of strategy if there is no understanding of the rules of the game and the aims and capabilities of the players. The bargaining between those who buy and those who sell reflects their changing expectations as they learn about each other's behavior and the factors that shape it.

A brief summary of the oil price "score" shows that in the period just before 1950 the per-barrel receipts or "take" realized by Saudi Arabia and Iran from oil extracted and exported by concessionaires was approximately 30 cents. From 1950 until the end of the 1960s these payments remained in the range of 70 to 90 cents per barrel, although a politically significant decline—that is, one which led precisely to the creation of OPEC—took place in 1959 and 1960. Since these figures correspond roughly to a 2-cents-per-gallon of raw material cost to the multinational companies that refine and market petroleum products, it is easy to see why the oil companies and their governmental allies in the Middle East chose to exploit the Saudi and Iranian sources of supply during the last two decades. There are vast oil reserves in other areas of the world; but since the cost of extraction is so low in the Middle East (currently about 10 cents per barrel compared to about $1.30 per barrel in the U.S.), oil companies and governments found it mutually profitable to obtain oil supplies there.[13] And the host countries found it extremely lucrative to be involved in a business in which they received five to ten times the cost of extraction.

During this period the relatively low retail price of refined petroleum products—combined with the economic recovery of Western Europe and Japan, and the general surge of consumption and industrial growth in the world—has resulted both in a vast increase in energy use and in an increased role for oil in satisfying that demand. In 1950 oil accounted for about 29 per cent of the 43 quadrillion Btu (British thermal units) of world energy use.[14] In 1968 this fraction stood at about 43 per cent of the world total of 64 quadrillion BTU. More than half the total increase in energy use during that time was derived from oil. More recent data for industrial nations, largest users of oil, show that the fraction of total energy obtained from oil in 1971 ranged from about 45 per cent in the U.S. to about 75 per cent for Japan. The OECD (Organization for Economic Cooperation and Development) average is roughly 62 per cent.

Of course, not all this increase in oil consumption is supplied by OPEC. Canada is virtually self-sufficient, and the U.S., by far the largest single user of oil, produces most of its oil needs domestically. From 1950 through 1973 the share of world oil output of the OPEC nations increased from about 30 per cent to more than 55 per cent.[15] In absolute terms there has been a tenfold increase in OPEC production to more than 30 million barrels per day.

During this period, the "take" price realized by Saudi Arabia is shown in Table 3. The $10.12 figure

TABLE 3
"Take" Price of Saudi Arabian Crude Oil

DATE	PRICE (U.S. DOLLARS/BARREL)
1950	.28
1951	.28
1952	N.A.
1953	N.A.
1954	.67
1955	.96
1956	.80
1957	.88
1958	.82
1959	.76
1960	.75
1961	.76
1962	.77
1963	.79
1964	.82
1965	.83
1966	.83
1967	.85
1968	.88
1969	.87
1970	.91
1971	1.07
January 1972	1.85
January 1973	2.30
June 1973	2.60
October 1973	3.70
January 1974	7.80
January 1975	10.12

SOURCE: M. A. Adelman, *The World Petroleum Market, op. cit.*, and Edward M. Bernstein, "The Economic Consequences of the High Cost of Oil," in U.S. Senate, *Implications of Recent OPEC Oil Price Increase, op. cit.* Data for 1950 to 1956 represent Aramco payments; Data for 1957 to 1969 are payments to Saudi Arabia.

reflects the take price announced at the Vienna meeting of OPEC in December 1974.

Now let us look behind the price changes and examine the most important of a series of economic and political confrontations and explore the often puzzling nature of relationships between key participants in the oil market.

These relationships can be seen as a kind of game. Essentially, any game consists of the interaction, over time, of players whose behavior influences the outcomes produced at various stages of the contest.[16] A vast range of situations may be portrayed in accordance with the possibilities for conflict and/or cooperation believed applicable to the relationships among any group of participants. A game-theory model, for example, is well suited for inquiring into that complex net of relationships among participants in the oil market which led to the U.S. foreign tax-credit arrangements that (since 1950) have redirected the tax revenues of principal oil companies from the U.S. Treasury to the governments of OPEC countries. According to the U.S. Tax Code, oil companies may subtract taxes to foreign governments from domestic tax payments. With this tax credit in hand, American oil companies were able to offer oil producers the amount they saved, thus in effect transferring abroad funds from the U.S. Treasury. The desire of the U.S. government to help protect American-based multinational oil companies against competition from British and French companies (acting in concert with their governments in trying to secure concessions in Saudi Arabia) was a key factor. Involved also in the rationale for this enormously costly measure

Energy and the Assault on Economic Order

was the U.S. desire to mollify Arab displeasure over the newly reestablished existence of Israel, and the still larger concern with securing oil supplies as part of the struggle between the "free world" and its Communist opposition.

Now, the game-theory approach calls for action to be studied in a dynamic context. As the game unfolds relationships between the players can change, and actions undertaken by each player or group of players may vary in accordance with changing circumstances and the changing beliefs held by players as to their options. The creation of OPEC, for example, came about as a response to the decrease in oil prices in 1959 and 1960. Although this was an event of obvious importance in creating the present dilemma, perhaps the most critical determinant of why the oil market has been transformed in the last five years is linked to the dynamics of the Libyan situation.[17]

The particular timing of Libyan oil finds, the post-1967 closure of the Suez Canal, the huge investments in Libya by "independent" U.S. oil companies, the Libyan revolution, and the interruption of oil flow in the trans-Syrian pipeline, all combined to precipitate the present prices. Libyan production was in full tide just after the 1967 war in the Middle East. At the same time guerrillas were attempting to sabotage the Syrian pipeline. The new revolutionary government of Colonel Qaddafi used this opportunity to raise its price 20 cents a barrel, threatening to expropriate American companies if its demands were not met. The United States oil community was divided on the issue—independent oil firms were willing to pay; major oil companies were not. In the end apparently 20 cents a barrel was considered not worth fighting about; the price increase stuck.

The ability of Libya to capitalize upon the now visible vulnerability of European consumers resulted in far more than a mere 20-cents-a-barrel price increase. Precisely because the oil market is a dynamic game, actions in it have two consequences: they produce both their own immediate tangible results and information. Information is the medium whereby knowledge is accumulated and expectations are shaped. Therefore, to describe the success of the 1970 Libyan price demands as "a flash of lightning in a summer sky" is hardly an exaggeration.[18] For they gave OPEC member states—and every other participant in the world's oil market—the sense of being confronted with a new reality, that they were, to follow out our metaphor, in a new ball game. Though it need not have been so, the lessons which each player appears to have learned have, through his subsequent actions, only served to reinforce and deepen this reality. Thus, "mere" pennies per barrel in 1970 became more than a thousand per cent increase by 1973.

There is no reason to suppose that the "winnings" of some players are composed automatically of the "losings" of others. Indeed, the outcome of economic and political processes is frequently other than "zero-sum." Many phenomena of economic growth and political cooperation produce situations in which, at least in absolute terms, all players benefit. Likewise, the result of such possibilities as prolonged embargoes and military action may be injury to all players.

Energy and the Assault on Economic Order

Although it is easy enough to speak of *the* oil game, there are in reality a myriad of interrelated games that collectively comprise the oil market. The market for oil is itself but one among a global ecology of economic and political "games." The deliberations and actions of those who play the oil game do not take place in isolation. They are linked inextricably to many other roles and activities undertaken by the various participants.

How each player views the relationship between his play in the oil game and his play in other games, and how he believes similar relationships describe the actions of other players, serve to establish his views about the stakes of the game and its rules. Depending upon a player's resources and outlook, he may conclude that the "true" stakes of the game include personal wealth, power, convenience; their national and corporate equivalents; environmental preservation, prospects for national development, the assertion of national and ideological claims, etc. Certainly it is not remarkable that different players, of different capabilities, will entertain and act upon their own differing beliefs as to the stakes of play in the oil game. But it is vitally important to recognize that over the course of time play may manifest some potentially catastrophic discontinuities in the rules by which each player believes himself and others to be bound. Because of the apocalyptic ramifications that these discrepancies may engender, to regard the oil market as other than one component (albeit an exceedingly important one) of the entire fabric of international relations would be both stupid and dangerous.

Who, then, are the players in the oil game? One

might suppose that few aspects of our problem could be as obvious or simple as identifying exactly who they are. But this is not so. Moreover, the nature of this question and our response is crucially important. For together they describe the boundaries of our vision and, perhaps more than any other factor, influence our recommendations for action.

Critical disparities among different classes of players hinge upon their perceptions of themselves and others regarding their:

1. Possession of oil and other energy resources, and other economic resources in general.

2. Oil needs, present and future.

3. Availability and price of alternative fuels and the feasibility of altering existing technological patterns to accommodate their use.

4. Need for the approval of other players. Of consummate political importance is the fact that all players face the need to produce results and/or policies to satisfy citizens, stockholders, hierarchical superiors, voters, allies, etc.

5. Legal status, e.g., possession of sovereignty.

6. Reaction time needed to initiate action and prepare for it, by developing necessary alliances and otherwise coordinating with other relevant players. Currently, the entire character of the oil game depends on the asymmetry of reaction time between OPEC governments and those of the major consuming nations.

There is, in effect, a cobra and mongoose confrontation in which the relatively sluggish maneuvering of the (otherwise formidable) consuming industrial nations is met by the fast reaction time of the sellers.

Energy and the Assault on Economic Order

Prices can be revised downward with ease and almost instantaneously to forestall what are taken to be credible threats of military intervention or even the introduction of new energy production technologies. The threat of price reduction may be sufficient to deter the gigantic investment programs that private entrepreneurs would have to undertake in developing a synthetic fuel industry or obtaining oil from shale and/or tar sands.

7. Extent to which information can be gathered about the needs, capacities, and intentions of other players.

8. Cultural and ideological attitudes toward:

    a. the sanctity of property rights and contractual obligations;

    b. the relative importance of long- vs. short-run benefits and costs;

    c. willingness to suffer and to impose suffering in order to obtain desired goals;

    d. the legitimacy of tactics such as the use of force, its threat, the use of bribes, deceit, innocent hostages, etc.;

    e. the appropriateness of selfish vs. altruistic play for each type of player; and

    f. wealth as opposed to power.

Each factor bears upon the issue of which game the various participants conceive themselves to be playing. Understanding these factors, therefore, is imperative. For the single greatest obstacle to analyzing the oil problem has been the number of different images of the "real" nature of the game projected by various participants and analysts. It should be clear, for example, that those who see the oil market exclusively as a financial concern could not explain the attempts of OPEC to buy into Lockheed and Grumman aircraft companies. Since both firms are abysmally in debt, it is obviously power rather than wealth (insofar as the two can be separated) which is the name of this game. Seen in this light, the apparent preoccupation of many commentators with the question of how best to "recycle" OPEC receipts corresponds to Schumpeter's caustic observation that "This is the way that the bourgeois mind works, always will work, even in sight of the hangman's rope." [19] If consumer nations refuse to recognize the political threat to their existence despite the abundant evidence before them, they must be prepared to swing in the breeze.

The major participants in the oil game often are identified primarily in terms of their membership in one of two presumedly competing camps—either as OPEC nations or consuming nations. This form of classification is a deceptive oversimplification. Neither camp is powerfully unified by anything much beyond their common linkage in trade in oil. To be sure, a Muslim heritage and antipathy toward Israel are shared by most OPEC members. But Algeria, Ecuador, Indonesia, Iran, Iraq, Kuwait, Libya, Nigeria, Qatar, Saudi Arabia, the United Arab Emirates, and Venezuela make up, to say the least, a mixed bag. They are divided in terms of ideology, wealth, population, and tactical inclination regarding the best way to utilize oil. The ability of major consuming nations, i.e., the U.S., Western Europe, and Japan, to exploit these differences, however, has been slight. In part this is because of even

Energy and the Assault on Economic Order

wider differences and competitive jealousies among the various consuming countries.

Historically the United States has played the central role in the world petroleum market. Until 1974 it was the world's largest single producer of crude oil; by a very wide margin it continues to be the largest consuming nation. What is more, only since World War II has the U.S. ceased to be the largest exporter of oil. Its role, however, involves far more than the data on geographic origins and destinations of oil flow suggest. In addition to the policies of governments, the actions of private organizations have in the past determined, and continue (although less dominantly) in the present to determine, the supply and price of oil. Since the turn of the century, American companies and, to a lesser extent, British and French firms have come to control virtually all phases of the oil market. Indeed, even during the last four years of radical rearrangement in the power relationships, the production-limiting institutions created by the oil companies apparently have been preserved by OPEC governments to serve their own interests.

In 1972 the seven major multinational oil companies accounted for 70 per cent of oil production outside the Communist nations.[20] In OPEC nations the relative role of these companies amounted to almost 80 per cent, as compared with their control of about 40 per cent of U.S., oil production. "Independent" oil companies accounted for the balance.

A distinguishing feature of oil company operation—both "major" and "independent," within the OPEC nations and without—is their remarkable cooperation.

They combine in various joint ventures and consortia to dominate production in Saudi Arabia, Kuwait, Iran, Iraq, and Abu Dhabi. They cooperate not only in extracting and transporting oil but in a variety of jointly owned undertakings in the related areas of coal, natural gas, and uranium production.[21]

It took no inspired genius to show oil companies that, by suitably coordinating their plans for production and marketing, they could reap the kind of profits theoretically available only under oligopolistic conditions. The basic requirement for accomplishing this is the ability to restrict production so that prices and profits will far exceed those realizeable under conditions of pure competition. This in turn is made possible only by the creation of effective institutions for enforcing the adherence of producers to agreed limits, and by the restriction of trading to prevent supplies from being offered by those who do not adhere to the quota system.

How an enforced scheme of restriction was instituted in the oil market is a rather complex story, and its existence long antedates the OPEC cartel. Since 1950 the growth rate of total oil production in OPEC countries has been invariant at about 9.6 per cent per year.[22] Despite vast fluctuations in the output of each individual nation, the total to be marketed by coordinating oil companies has grown constantly in accordance with industry "forecasts." Although the relative role of the companies vis-à-vis OPEC governments is changing, the existing average program quantity (APQ) system for establishing production quotas is apparently still in operation.

Energy and the Assault on Economic Order

Arrangements of this kind are vulnerable to significant amounts of uncontrolled output. So far, however, it seems that all participants on the producer side have been unwilling to jeopardize their presently lucrative positions by trying to introduce additional supplies. In the domestic U.S. oil market oil companies and government officials have shown a close and often reciprocal concern for one another's needs. Thus the Connally "Hot Oil" Act protects companies against the interstate marketing of oil not covered by existing (and price-raising) quota agreements to limit the amount of domestic production.

The impact of the U.S.S.R., China, and Eastern European nations on the world oil market is slight, at least as far as their being buyers and sellers is concerned. Together they produce about 10 per cent of the world's output. At present the U.S.S.R. exports some oil.[23] However, as the Russians' internal energy requirements continue to expand,[24] evidence suggests that this surplus will have been only a transitory phase.

Although their significance is far from obscure, there are several types of players whose role is discounted or overlooked altogether. This is generally because such players lack an organizational identity and must be referred to in such vague terms as "interest groups" or, still more nebulous, "interests." The role played by "consumers," and expectations about their possible political reaction to stringent gasoline conservation measures, for example, looms large as a factor shaping the play of virtually all participants. Less significant, but still influential insofar as it constrains the planning and behavior of participants, is the part played by opponents within their own countries of the OPEC regimes.

Where possibilities exist for forming or dissolving a variety of alliances among different groups of players; where the strength of alliances exerts a major influence on the bargaining; and where the potential for mutual benefit and/or mutual loss is significant, game theorists have delved into the sometimes paradoxical requirements of rationality. What is rational behavior for each of several players when they act in concert can equally be viewed as the height of irrationality when the same behavior is displayed by the same players conceived to be acting alone.

A further complication arises from the subtle connection between time horizons and notions of rationality. Players may be willing to suffer losses—or at the very least proclaim their willingness to do so—if they believe it will "teach" other players that their own interests will suffer also. This is precisely the point that Sheikh Yamani of Saudi Arabia is making when he states that OPEC countries would not be hurt by a 10 per cent cutback in oil consumption since they are able to withstand a 33 per cent reduction.[25] Obviously the willingness of various players to suffer the consequences of decreased or even totally interrupted trade in oil depends upon the extent of their self-sufficiency, their reserves, and their readiness to alter styles of living. Which ones, for instance, are best able to impose sacrifices on their own people? For the purposes of bargaining about oil, the key fact about such threats as Yamani's is their believability. Moves in the oil game frequently are cast in the form of such threats: *viz.* the

Energy and the Assault on Economic Order

suggestion by Kissinger that military intervention is a possibility being studied by the U.S.

As the history of the Standard Oil Company makes clear, players may choose to sustain temporary losses if by doing so they can force rivals out of the game entirely and thus reap greater long-run profits. Game theory cannot offer a specific answer as to when such self-inflicted reversals or the threat of mutual harm will produce long-run benefit. This always depends upon how other players choose to respond.

Moving from the theory to the actual play of this particular game, one of the factors that have frustrated the efforts of many people to understand events in the world oil market stems from the frequently perplexing nature of the links between oil companies and the governments with which they deal. Although we can hardly promise to make clear all or even most of the intricate details of those relationships, by explaining the most significant ones we can help to dissolve some major mysteries surrounding oil market behavior.

Two very obvious facts must be recognized from the outset. First, whatever the political power of the companies, their legal status is an inferior one to that of governments. In their dealings with both sellers and buyers they are subject to the laws which sovereign states choose to impose upon them. Second, the realities of economic life are such that both cooperation and competition operate in their relations with governments as well as in their relations with other oil companies. Thus it would be erroneous to presume an identity of interest between the oil marketing firms and either their own national governments or the governments whose crude oil they refine and distribute. In certain situations, the multinational oil firms may be constrained to act as if they were the representatives of importing nations or spokesmen for the consumers in those nations. Conversely, circumstances may cause them to act on behalf of OPEC governments believing their best corporate interests to coincide with those of the exporters. The only general rule that can be laid down flatly is that oil companies and governments can be counted on to try to use one another in accordance with their respective conceptions of national and corporate interest. To the extent possible, each of the various types of participants in the world oil market tries to alter the institutions and conditions of that market to maximize its own benefit. Although both exporting and importing governments have at times in the past acted as though they believed their own interests to be best served by some "invisible hand" guiding the actions of the multinational oil firms, it is now apparent that reliance upon such beliefs can arise only from some mixture of political impotence and/or self-deception.

Historically the United States has been the most influential actor in shaping the relations between governments and companies in the world oil market. Five of the seven largest multinational "majors" are U.S. firms and for the past five decades at least—from the time we were the world's largest oil exporter to the present, when we are the largest importer—U.S. oil policy has been a prominent issue. While focusing as much on how to regulate the domestic activity of oil companies as on the question of their foreign operations, the de-

Energy and the Assault on Economic Order

bate on oil has centrally featured the requirements of national security. It is a widely held view that in both the First and Second World Wars the allied powers "floated to victory on a sea of oil." [26] Thus U.S. policy has long been aimed at insuring the existence of plentiful domestic oil reserves. While some lands were set aside and designated as strategic reserves, the main thrust of governmental action has been to seek to encourage private exploration. To this end, it has generally been government policy to make the oil business a profitable one. The substantial oil depletion allowance and provision for the rapid expensing of intangible exploration costs are prominent legislative contributions to oil industry profitability by means of decreased federal taxation.[27] Similarly, such things as the Connally Act, federal tolerance of state-imposed production quotas, the activity of the Texas Railroad Commission, and the oil import quotas established by the Eisenhower administration have all served to keep profits high, and thereby, it is argued, furthered America's national security.[28] Those who have maintained the need for direct federal action to provide the necessary petroleum found few allies in the industry, the military, or elsewhere.

Being able to exploit the vast and cheaply tapped resources of the Persian Gulf States was thus seen as serving a confluence of interests: the oil companies, the military, and consumers as well would all be the beneficiaries. Therefore, at the initiative of the National Security Council provision was made to have oil company concessionaires in the Gulf area transfer their liability for U.S. federal income tax on foreign profits directly to the governments of exporting countries. However, to qualify for the foreign tax credit, the oil companies and the host countries had to cooperate in the building of something that resembled a locally imposed income tax on their foreign operations. In monetary terms the foreign tax credit on oil is immense, dwarfing the impact of the depletion allowance, for example. Its current annual cost to the U.S. is many billions of dollars.

Obviously the transfer of oil company payments away from the U.S. Treasury and toward the OPEC nations is beneficial to the latter. Presumably it is also a matter of indifference to the companies where their taxes go. However, because these firms are vertically integrated and because their marketing competitors in consuming nations do not possess adequate supplies of crude oil of their own, it is most advantageous to establish transfer prices that shift all corporate profit to their overseas producing operations. Thus the current system of taxation is doubly beneficial to producer nations. How is it to be explained not only that this system was put into operation by U.S. design in the first place but that it continues with U.S. collaboration—a situation which could easily be remedied by a unilateral modification of our tax laws? The question is not easy to answer. Nor are similar questions about why the U.S. government allowed (or recommended) a separate and price-increasing settlement in the Teheran-Tripoli negotiations in 1971, why it has chosen to acquiesce in oil company administration of the 1973 OPEC embargo, or why it continues to permit oil companies to refuse to furnish data on their costs of operation.[29]

Energy and the Assault on Economic Order

Some have sought to answer such questions in terms of conspiracies between policy makers and oil company executives. Interestingly, but perhaps not surprisingly, the Shah of Iran and other OPEC spokesmen encourage the common view that it is the oil companies who are the "bad guys." For others, current U.S. governmental action—or rather, inaction—is little more than a necessary acknowledgement of our presently powerless position in the oil game: we may offer understanding, but have not much assistance to give to the companies in their hour of tribulation. One particularly fascinating hypothesis is that all the actions which have led to the current price levels were instituted by the deliberate and concerted action of the companies, the U.S., and OPEC! [30] According to the proponents of this view, such concerted action springs on the one side from the desire for profits by OPEC and the major oil companies and on the other, from the desire of the U.S. and these companies to destroy their more dependent and therefore more vulnerable rivals. This hypothesis would surely account for a number of hitherto seemingly incomprehensible pronouncements to the effect that OPEC had best respect the vulnerability of consumer nations!

Adelman, however, suggests that there has been no U.S. government policy either vis-à-vis the companies or for that matter, with respect to the entire crisis.[31] Perhaps the most intriguing and baffling insight into the relationship between the U.S. government and the oil companies is to be found in the testimony of John J. McCloy. In response to a question about whether it might not be desirable for the U.S. now to take the unprecedented step of entering directly into the bargaining process, McCloy's answer is that such a step, by making political issues predominant, might lead to "confrontations." [32] In effect, he is saying that the best way to attend to this country's political interests in the oil market is by simply letting the results of some presumably non-political bargaining between OPEC governments and oil companies determine what is to happen. The fear that "confrontation" may result from governmental participation on the part of both buyer and seller nations follows from a preoccupation with securing agreements almost irrespective of what these agreements actually are. Whatever reservations one might have about either the motives or the competence of government representatives to serve the best interest of government, it is simply preposterous to think that somehow or other we, as a nation, can obtain better results by putting our trust in people who are obliged to pursue the aims of the oil companies that employ them.

The fact that oil companies are permitted to market products within the consumer nations is the main source of their current value to the OPEC countries. These companies enjoy a high degree of legitimacy precisely by virtue of their being concerned with their own rather than with any governmental interests. If it is they, rather than the U.S. government, who pay taxes to foreign nations, why should OPEC mind? Although textbooks on economic warfare tell us that an ideal avenue of penetration into the economic life of one's enemies, present or prospective, is through exploiting the divergence of interests between the private firms and

Energy and the Assault on Economic Order

governments of those economies, why should anyone now start to quibble about the fact that the U.S. government is currently unrepresented in the dealings of the oil companies with OPEC? [33] The answer, of course, lies in a survey of the results. There is just too great a difference between the needs of the United States and those of the oil companies to expect, or even to ask, that the latter abandon the interests of their stockholders in order to undertake the responsibilities that our government has shirked.

The common interest shared by oil companies and exporting nations in the crude oil production and refining facilities they jointly control is so gigantic and so much at the mercy of those nations that, for example, we should expect the oil companies to oppose any alteration in their present role as "tax collectors for OPEC." [34] It would be unnatural in the extreme for the companies to do anything whatever to antagonize the exporting nations, for the profits of accommodation are as large as their tenure is precarious. Both companies and exporter governments know that the yearly returns in profit on invested capital to the companies exceed 100 per cent. Thus the expropriation of their assets, even if financial compensation were to be offered, would be a grievous blow.

Among the profusion of analogies that have been put forth to further our understanding of the oil problem, the most useful one is the analogy between the oil game and the narcotics market. The actual mechanisms of the addiction to oil are diverse. They range from the truly addictive human dependence upon food (whose production in the necessary quantity depends upon petroleum to provide energy for fertilizers, farm machinery, and transportation) to the less compelling psychological dependency upon maintaining existing lifestyles, particularly those associated with the automobile. In each situation, the ability of addicts to respond to changes in the "street price," or the availability of oil, is a function of the technological alternatives available, their will and capacity to alter technologies and life-styles, and whether or not they possess the wealth required to invest in creating the new reality.

The markets in opium and oil have a fascinating similarity. The trade in both is international, and along the way from producer to consumer, or addict, there are several key figures. If we take consumers as addicts, the oil companies as pushers, and the producing nations as the growers, or at least those who control the land on which the crop is produced, we can begin to understand the whole process better. The transactions involved before the final product is consumed involve a good deal more than simply a seller and a buyer. The raw material must first be obtained, by an investment of time and money; the same is true of refining, transportation, and the other details of distribution. Each stage in the process is vulnerable to interference by governments or private groups claiming that the participants in the market need their "protection." Occasionally this protection may take a form whose legitimacy is accepted—a demand for royalties and tax payments, say, imposed by those with a recognized sovereign right to do so. In other instances, the price of protection will consist of payments delivered covertly

Energy and the Assault on Economic Order

to public officials, terrorists, criminal syndicates, etc.

The recent behavior of the world oil market can be seen as the result first of one and then of all the growers having realized that, in view of the likely inadequate coordination of refiners, pushers, and users, they had a chance to increase their profits drastically. They have achieved their end by taking a far greater share of the increased revenue that oil marketing companies can extract from consumers.

Amuzegar's description of this phenomenon is most artfully put.[35] In the theoretical language of economics he observes that marketing companies are now collecting more of the "consumer surplus" (the additional amount buyers would have been willing to pay if market prices had been higher), while the exporting countries are in a position to gather in the huge Ricardian rents the oil companies are now able to pay. In effect, growers have forced pushers to work for them instead of the other way around, in the process, to charge addicts a much higher price—though a price the users are apparently still able to bear. The growers had to convince the various pusher organizations that protection from the governments which had historically stood behind them was now ineffectual. This they did, and most skillfully, by playing off the competitive interests of the oil companies the governments of their respective national origins. Indeed, OPEC has been spectacularly successful in exploiting both greed and fear. Many participants in the market, for instance, were initially unable to calculate whether they were being helped or hurt by OPEC's actions; even now some of them still find this hard to figure out. Clearly, U.S.-based multinational oil companies have profited. Others, such as the U.S. government, have also realized advantages, at least relatively, vis-à-vis traditional trade rivals who must support a greater oil import "habit." But even when they are real, such relative advantages to the U.S. government may be only temporary, for gains like this are won at the cost of ultimate absolute losses to all non-OPEC nations. Likewise, the advantages currently being enjoyed by the "pushing" companies may be no more than a short-term capitalization of long-run assets in the throes of de facto expropriation.

We can sympathize with those addicts and their governments who, unable to meet the stiff new final prices, can derive a perverse satisfaction from their own misery by watching their previously mighty pushers suffer. Unhappily, such satisfactions offer cold comfort to those whose food and work depend for their very existence upon the availability of oil.

Odell and Adelman, for example—two of the most astute observers of the oil market—regard the links between governments and multinational oil companies as a key factor in explaining our current predicament.[36] While they differ in their estimates of the relative strength of the U.S. government and U.S. based multinational oil companies in their efforts to use one another, they see this tie between the "pushers" and the governmental "syndicate" that controls this sales territory as being of paramount importance. Odell subscribes to the view that the U.S. government has acted consistently on behalf of U.S. oil company interests and that, despite shedding some crocodile tears, it is con-

Energy and the Assault on Economic Order

tent with the continuing dominance of American as opposed to Western European pushers. Adelman, on the other hand, comments that U.S. governmental policy has been anything but consistent and, more important, that this relationship is the primary instrument by which OPEC governments are able to extract all that consumers in the U.S. are "willing" to pay.

The narcotics analogy appears even more apt as Adelman points out that even where governments may choose to pursue the interests of their domestic consumers (addicts) they cannot, using present bargaining methods, seek out and secure a reliable "connection" except at a cost so prohibitive as to make the entire enterprise counterproductive.[37] Indeed, when a less than dominant buyer or seller tries to insulate himself against the uncertainties of the marketplace, he will be forced to pay a premium in the form of higher prices or to accept a lower price than he might otherwise have secured by refusing to commit himself in advance.

Another value of the narcotics analogy is to remind us that denial of the "right" to sell opium has already figured prominently in one war. It is possible that the use of sovereign rights to prevent the sale as well as the extraction of oil is an option that may be exercised by players other than OPEC governments, an alternative, among others, whose potential import we must examine.

The narcotics analogy spotlights the conflicting requirements of consumer sovereignty and national sovereignty. In most situations, of course, there has been no need to reconcile these principles. Where conflicts between the two are evident, our collective political needs have taken precedence over the "needs" of the marketplace.

The mere fact that there are potential buyers and sellers willing to trade in such commodities and services as narcotics or prostitution is not sufficient to justify their legalization when doing so is believed to be destructive of the entire community. It is precisely this notion that underlies prohibition of the sale of "strategic" materials such as sophisticated electronic circuitry and ultrafine machine tools. As yet, though, the growing recognition of just how pernicious the massive outflow of capital has been to our society is unmatched by a serious determination to check the trade from which it springs.

This failure is one of two dominant factors in shaping the strategy of players in the oil game. The unwillingness of the U.S. government and others to frustrate the petroleum-consuming desires of their citizenry is obvious to all. The source of this unwillingness is political, as is the source the one-sided and self-imposed refusal to exert, or even to publicly debate, the use of force. It is by no means certain that a demonstration of political will, which would be signaled by a change in policy regarding either or both of these factors, can alter the so-called purely "economic" results of the game. However, it is absolutely certain that the absence of such changes will keep our national destiny and that of other consuming states in thrall to the sovereign consumer addict.

To speak of oil as analogous with narcotics suggests the possible irreversible nature of the processes we must deal with here. Akins goes even further when he speaks of our predicament as one of "how to get the genie back in the bottle." [38] The question, then, is whether recent oil market developments are reversible—and if so, how.

The first thing to say is that it would be most useful to set aside the entire question of who is justified and who is not justified in the present situation. What is needed rather is a perspective on the future and the future consequences of various arrangements. Admittedly, this approach offers little emotional satisfaction to those who want to see the wicked punished. But for the utilitarian, the only villainies relevant to his consideration are those that might occur in the future, and his most pressing obligation is to learn how to avoid them.

We cannot, however, predict future consequences without a solid concrete sense of what has happened in the past. Whether or not we have in the present crisis entered into an irreversible process of redistributing the wealth among nations, and what such a redistribution might mean, are questions that can only be examined on the basis of our knowing what has already been done by those nations who now promise, or threaten, to control the world's wealth. We must, for example, consider the available evidence on the comparative distribution of wealth in the consuming countries and in the disparate collection of political entities which make up OPEC. Also, we must examine the likely courses of action to be pursued by the political regimes of the OPEC countries, and by other private accumulators of wealth within these regimes, as they use the surplus funds at their disposal.

The Great Détente Disaster

Quite clearly, economic growth and future economic wellbeing depend upon investment in productive capacity. If those who control the wealth of societies choose to spend it all upon the consumption of goods and services or to "invest" it in ways that do not enhance future productive capacity, real wealth will decline. From the standpoint of the individuals making those expenditures, the hoarding of consumer goods, the purchase of military hardware, short-term deposits in foreign banks, etc., may be quite rational, but the overall effect of their decisions will nevertheless be a harmful one. The wealth of those who hope to profit may be enhanced; but such enhancement of their wealth can be accomplished only by imposing a still greater cost upon the entire economic system. Further, if the extent of such nonproductive uses of funds is large enough, there may be a worldwide systemic collapse in which everybody, including the short-term winners, loses.

The very sticky problem of the link between economic development, wealth distribution, and irreversibility is clarified in Keynes's account of pre-World War I Europe:

> Europe was so organised socially and economically as to secure the maximum accumulation of capital. While there was some continuous improvement in the daily conditions of the mass of the population, society was so framed as to throw a great part of the increased income into the control of the class least likely to consume it. The new rich of the nineteenth century were not brought up to large expenditures, and preferred the power which investment gave them to the pleasures of immediate consumption. In fact, it was precisely the *in-*

equality of the distribution of wealth which made possible those vast accumulations of fixed wealth and of capital improvements which distinguished that age from all others. Herein lay, in fact, the main justification of the capitalist system. If the rich had spent their new wealth on their own enjoyments, the world would long ago have found such a régime intolerable. But like bees they saved and accumulated, not less to the advantage of the whole community because they themselves held narrower ends in prospect.

The immense accumulations of fixed capital which, to the great benefit of mankind, were built up during the half century before the war, could never have come about in a society where wealth was divided equitably. The railways of the world, which that age built as a monument to posterity, were, not less than the pyramids of Egypt, the work of labour which was not free to consume in immediate enjoyment the full equivalent of its efforts.

Thus this remarkable system depended for its growth on a double bluff or deception. On the one hand the labouring classes accepted from ignorance or powerlessness, or were compelled, persuaded, or cajoled by custom, convention, authority, and the well-established order of society into accepting, a situation in which they could call their own very little of the cake that they and nature and the capitalist classes were co-operating to produce. . . . I seek only to point out that the principle of accumulation based on inequality was a vital part of the pre-war order of society and of progress as we then understood it, and to emphasise that this principle depended on unstable psychological conditions, which it may be impossible to re-create. It was not natural for a population, of whom so few enjoyed the comforts of life, to accumulate so hugely. The war has disclosed the possibility of consumption to all and the vanity of abstinence to many. Thus the bluff is discovered; the labouring classes may be no longer willing to forgo so largely, and the capitalist classes, no longer confident of the future, may seek to enjoy more fully their liberties of consumption so long as they last, and thus precipitate the hour of their confiscation.[39]

For us, as for Keynes, systemic change is the main concern. Although both situations involve the interaction of domestic and international redistributions of wealth, the latter are now clearly more pronounced. It would be misleading to accept the frequently proclaimed idea that our present crisis is causing, and is caused by, a redistribution of wealth from the rich to the poor. Certainly, on a national basis, wealth is flowing toward the members of OPEC at an historically unprecedented rate. This wealth originates primarily in the industrial nations of the West, but a substantial flow is drained from those least able to afford it—the non-OPEC nations of the so-called Third World.

Because so much concern has crystallized around the purely national impact of the present transfer of wealth, it has been easy to overlook other even more significant characteristics of this transfer. The essential feature of the new redistribution is that it transfers wealth to the OPEC governments from the middle and working classes in industrial nations, and from those who control the meager financial resources of the oil-importing Third World. Added to this the existing wealth distribution patterns within the OPEC countries, it is quite likely that the wealth is actually being transferred from the "poor" to the "rich" rather than the other way around.

If we choose to put considerations of equity aside,

then in accordance with Keynes's view we need not rush to the conclusion that this anti-egalitarian redistribution is by itself a bad thing: whether it is good or bad must ultimately depend upon the uses to which the wealth is put by its new owners. If we find the OPEC regimes willing to commit their surplus funds to long-term investments—needed to increase the world's productive capacity—the redistribution may very well prove beneficial. On the other hand, if the surplus funds are spent on luxuries, armaments, and other productively "useless" consumption, and on the purchase of highly liquid assets, this redistribution will have a catastrophic impact. The present behavior of OPEC governments, insofar as it can be discerned, suggests that they are setting themselves upon the latter course.

Keynes argues that the trend toward a more egalitarian pattern of wealth distribution is irreversible and "natural." Basically he regards this as stemming from the unwillingness of people to delay gratification. His view is no less political than psychological. For the actual willingness of humans to delay gratification quite clearly depends upon whether they have the means, and the institutions, that permit them to resist such delays. In our present situation, there are no encompassing political structures through which those whose wealth is being drawn off can translate their opposition into an electoral or any other kind of force capable of stopping the chain of events. Particularly in the industrial nations, a growing refusal to tolerate what most people have come to regard as the natural and irreversible trend toward increased wealth may find expression in a demand for natural military action.

The Great Détente Disaster

There is a mutually reenforcing interaction between OPEC preferences for armaments and highly liquid forms of investment and the threats of military action made by consuming nations. OPEC and the oil importers react in very natural ways to what they justifiably regard as each other's menacing behavior. And in so doing they contribute to each other's reasons for pursuing a mutually destructive course. In the case of arms sales this interaction is particularly obvious. As the industrial states seek to finance oil imports by selling weapons, not only do they raise the level of future conflict, if it should occur, but they absorb the OPEC funds which might otherwise have financed more economically productive activities. While the reactions of OPEC and the oil importers are understandable, as well as natural, they exacerbate the trend towards systemic collapse.

To this point in time an important stabilizing force has been the fact that suffering, as the consequence of high oil prices, has been concentrated in the most impotent consumer nations. However, with the passage of time, systemic effects will cause even the most powerful and economically insulated states to be hurt. Moreover, as a recognition of this eventuality spreads, its implication becomes clarified. All may come to believe that the basic choice before the oil importing states is not whether they must resort to military action, but rather to determine the form and timing of that action.

Although the recent docility of the industrial states might lead some to conclude that they are "paper tigers," a more realistic appraisal is necessary. As eco-

Energy and the Assault on Economic Order

nomic stability is perceived to deteriorate, the precarious dominance of the OPEC states may be likened more accurately to that of the lady in the familiar limerick that concludes:

> They returned from the ride
> With the lady inside
> And the smile on the face of the tiger.

Therefore, the overriding strategic principle for OPEC players in the oil game is to insure that all players continue to regard the game exclusively in economic terms. In this respect, an oil embargo is a particularly dangerous and unpredictable weapon. For not only its actual use, but the very threat of that use, will cause consumer nations to view the oil game as primarily political. Evidently it was a desire to forestall the consequences of such altered perceptions that led the Shah of Iran to recently proclaim that he would refuse to join in any futute embargo.

To put the matter in somewhat different terms OPEC players need to maintain the separation between economic and political dimensions of the oil game in order to preserve stability and thus the very existence of the game they are winning. From OPEC's standpoint, the economic goal of profit maximization must be understood as relative to a proper time context. And we must recognize that variations in the "understanding" of the appropriate time horizon that each member's resources and political objectives lead them to will be a major source of potential instability in the OPEC alliance.

What, then, is to be done? To achieve our basic goal of preserving the integrity of the world economic system and our highly favorable position in that system, we must bring about a more favorable relationship between the supply of oil and demands for it. We may work toward that goal by increasing supply and/or decreasing demand. In general terms, supply can be increased by developing new fuel resources, substituting other fuels for petroleum, and improving the efficiency of energy conversion devices; and perhaps more important, demand can be decreased by altering patterns of behavior so as to conserve the oil that must be used.

But increased supply and/or decreased demand are only part of the story. For neither the price nor the availability of oil in the world market is wholly dependent upon the physical aspects of its sources and uses. They serve only as rough boundaries that limit the very wide freedom of action available to all players in the oil game. An extensive range of possibilities regarding the details of availability and price exists at all times. Actual market conditions reflect the less tangible realities of bargaining which, in turn, rest upon the participants' perceptions and expectations. Therefore, we must be careful to avoid presumptions based on our beliefs about the properties of purely competitive markets. As both Adelman and Sheikh Yamani—the Saudi oil minister—make clear, because the world oil market deviates so far from pure competition there is no reason whatever to suppose that a decrease in demand or an increase in oil availability for export must result in a lower price.[40]

We may usefully distinguish, therefore, between policy options which are aimed at making it possible for us to endure current price levels and policy options which are designed to change these price levels. To accomplish the former we require an arrangement that will persuade those who possess crude oil to export it to those who wish to buy it. Of course, in the marketplace, "wishing" is not enough. Desires must be backed up by a quid pro quo that the prospective seller is willing to accept. Given their present dwindling financial reserves, and their dwindling ability to earn additional reserves via trade, virtually all the oil-consuming nations will soon be unable to pay for oil imports. At current price levels, trade in oil will continue if, and only if, either the sellers are willing to grant credit or buyers can borrow funds elsewhere. The funds paid out for oil can of course find their way back into the hands of consumer nations in a variety of ways: as payment for arms shipments and the purchase of other commodities produced in the consumer nations, and as "recycled" surplus funds that OPEC members choose to invest in consumer nations. These investments may be direct, involving the purchase of foreign assets or bilateral loans to governments. They may be indirect and channeled through intermediary institutions such as the Euro-dollar banking network, the World Bank, or the International Monetary Fund. Now, recycling of this kind may be both necessary and desirable; nevertheless, it is difficult to imagine that the pattern of OPEC investments will reflect anything other than the self-interest of OPEC countries as they choose to conceive it. OPEC may make modest contributions to some sort of financial "safety net" which helps consumer nations with their immediate short-run problems; but these stop-gaps may not be translatable into the kind of long-run capital investments which will—*pace* Keynes—be necessary for the long run. Moreover, the problem is more even than the lack of capital investment needed to enhance, or at least maintain, worldwide levels of real production. Poorer countries need to finance current *consumption*. It very much remains to be seen whether there will be recycled funds available to make viable the continued existence of Third World countries, whose main claim to creditworthiness resides in the humanitarian aspirations of their creditors.

Another way to sustain that viability would be to set up a two-tier system of prices, for oil and perhaps for other commodities such as food. Under a two-tier price schedule, the price of items sold to poorer nations would be lower than of those bought by others. Senator Jackson, for example, has proposed a plan under which unused OPEC production capacity would be dedicated to the production of oil to be sold at cost (roughly 10 cents per barrel) to poor nations.[41] Despite their general appeal on grounds of equity, two-tier plans must overcome at least two major obstacles. First, their administration requires considerable monitoring and a serious effort at enforcement in order to avert profiteering by the surreptitious transshipment of cargoes from poorer to richer nations. Second, the setting of prices would itself remain as a critical and complex issue, for the specific price levels selected would determine how the burden of supporting the poor would be shared between exporters and their fully industrialized customers. In other words, would prices on the higher tier be set so high as to include the total burden of compensation for the low prices on the second tier? OPEC members will be not anxious to forego the profits they would otherwise have made, nor, on the other hand, will the "richer" importers fancy adding yet a further self-inflicted burden to the load under which they are already staggering.

Oil imports that cannot be financed out of current trade may, if sellers are so inclined, be purchased for promises of future payment and/or the transfer of equity rights. A favorite theme of many so-called "responsible" observers is that really very little choice rests with the sellers: whatever they might think or believe, in the end they simply have to recycle funds in a beneficial way. This, alas, is plainly not true. Although long-range investment opportunities may be regarded as desirable and be seized upon by OPEC, there is no compulsion for them to do so. Even where OPEC members are willing to undertake such investments, nothing constrains them to distribute their investments in accordance with the financial needs and the desire to buy oil of their customers. When prospective buyers lack all other means to pay, sales may simply not be made. Indeed, retaining oil in the ground may be viewed as more attractive than various investment alternatives.

OPEC investors are likely to face an assortment of "Quaker Oats" policies—that is, policies which attempt to restrict the areas of the economy in which they may invest large quantities of money. Designed to avert political penetration in the form of alien control over vital national industries, such policies for understandable reasons have already begun to limit the range of opportunities open to OPEC. But there are even more powerful deterrents to OPEC investments. First and foremost, such investments would be highly vulnerable. Inflationary policies, currency manipulations, blockages of payment, threats of expropriation, and expropriations themselves could at some future time be used by borrowers should the circumstances arise that altered their dependency on oil imports. This is, of course, a possibility well understood by both OPEC and the consuming nations. Certainly, OPEC would have to exercise great caution in the making of any serious commitment to long-term investments, for they are well aware that deceit might be resorted to against their interests. Even if no such deceit is currently contemplated by the industrial world, changing circumstances—particularly those brought about by a recalcitrant public, and stemming from a systemic collapse or a great breakthrough in tapping new sources of energy—might result in the loss of their investments.

The basic problem in dealing with strategies for endurance is the need to ascertain their political feasibility. Thus, Chenery's assertion that, if the transient initial shocks to the world economic order can be mastered, adaptation to high oil prices is possible, is clearly true.[42] The issue, however, is not whether the institutional rearrangements he envisions are possible but rather whether they are sufficiently attractive to command the support they require. Whatever satisfaction we derive from recognizing that things could be tolerable if only conditions in the oil market were stabilized, and if only OPEC undertook the financial responsibility of launching a latter-day Marshall Plan, etc., we

Energy and the Assault on Economic Order

cannot mistake our fantasies for reality. The prospective returns that OPEC nations might realize by assenting to Chenery's proposals are most unlikely to be regarded by those nations as being the most attractive of the options at their disposal. Implementation would be uncertain. And the dependence upon the future behavior of non-OPEC regimes that such a restructuring of the world's economy would impose upon OPEC members constitutes a most serious and, in our view, a virtually insurmountable political obstacle to its adoption.

As far as the rise in oil prices itself is concerned, we do well to remind ourselves that it was not occasioned by the fortuitous workings of some law of nature. It resulted, after all, from choices made by men. Although these choices were grounded in particular people's perceptions of the past and present, and their sense of a possible future, they could also have produced other price levels. In any case, given the current price of oil, it is problematic whether even our most ingenious efforts would be sufficient to insure the survival of the world's economic system as we now know it. There is, however, no need for us to discover some miraculous new means for getting along on the basis of a destructively expensive source of energy. We have it within our ordinary present power to alter the situation so that a price decrease results. Perhaps the most basic component of a strategy for bringing about a reduction in the price of oil lies in the area of energy conservation. The producers and marketers have seen demonstrated over the past four years not only how inelastic is the demand for their product but just how timid the governments of consumer nations are. In the United States, for example, the main governmental contribution to the conserving of oil has been a mild suggestion or two. The one enforced conservation measure, the 55 mph speed limit, is helpful—but only a drop in the bucket. An all-out program should be mounted to shift passenger travel from private autos toward mass transit systems. Where this cannot be done, car-pooling must be encouraged by strong inducements and/or the use of negative sanctions. Drastic improvements in the present fuel economy of cars must be brought about. Other policies with great potential would be shifting intercity freight transportation away from trucking and rapidly developing our vast coal resources.

A variety of techniques may be used to reduce consumption by making petroleum products more expensive and/or less accessible domestically. The recently announced intention of creating a large crude oil stockpile would certainly be a step in the right direction. Almost irrespective of the results of particular conservation policies in the saving of fuel we will benefit by signaling the termination of our present role as helpless addicts.

It is vital to the solution of our overall problem that institutional factors be made to work for us rather than against us. The first and most important action we must take toward solving the present crisis is to recognize its full extent and significance. We must be willing to face the very unpleasant fact that our present predicament is the result of past wishful thinking and almost unwavering refusal to "see" reality. We have been careless in our treatment of oil companies. We have permitted—

Energy and the Assault on Economic Order

even fostered—a situation in which the oil companies, perhaps contrary to their own wishes, actually serve the interests of exporting countries and oppose those of the United States and oil-importing nations in general. They serve as tax collectors for the OPEC states by using the foreign tax credit provisions to siphon gigantic sums from the U.S. Treasury. They operate virtually as marketing agents, extracting for their OPEC principals all the market will bear. Worse still, present relationships between the oil companies and the U.S. government make it virtually impossible to implement corrective action by improving the domestic supply situation. Because these companies are conceived to be legitimately and exclusively concerned with their own financial self-interest, we cannot direct an extension of their exploration activities, nor can we require their development of synthetic fuel production capabilities and plants for extracting oil from shale or tar sands. Instead the government sees itself capable only of influencing oil company behavior by offering financial inducements (in the form of natural gas decontrol, for instance), by underwriting the appreciable risks involved in the development of new fuel technology, and by allowing—rather than requiring—the exploration of new lands. The relationship between the U.S. government and these companies has become one of virtual producer sovereignty. Nowhere is this more clearly shown than in the willingness of U.S. oil companies to provide detailed data on their own deliveries and third-party deliveries of petroleum to U.S. military forces to assure enforcement of the 1973 embargo. Inexplicably the oil companies have been imagined to represent the consumers' interests in their bargaining with OPEC. Their response has been predictable in such dealings; here, OPEC governments wield the formidable stick of threatened expropriation while the U.S. government and consumers have only carrots at their disposal. The economic self-interest of the oil companies has led to their playing ball with their overseas suppliers. Some critics even go so far as to blame the major oil companies for teaching OPEC the fundamentals of establishing a cartel. This view is unduly arrogant but does at least make the origins of the cartel clear.

It would be totally unrealistic to imagine that the relative roles of the U.S. government and the oil companies can be altered to the point where the latter are conceived, and conceive themselves, to be primarily interested in promoting the national interest. Moreover, such a change would not be desirable. But some aspects of the oil market can be changed so that if the companies do not work for the national interest at least they will not work against it. In this respect Adelman's suggestion that the U.S. government become the sole importer of oil to this country seems worthy of adoption.[43] We doubt that the provision for secret bidding by sellers would be sufficient to induce individual cartel members to "chisel" on the cartel agreement. However, we will at least be able to prevent the kind of co-operation between producers and distributors which is detrimental to us. There is no way that OPEC countries can capitalize on threats to expropriate the assets of the oil companies if they are prohibited from dealing with them. In simple terms, the assets could not be used as a hostage to force the companies to cooperate in extracting money from consumers.

Beyond this, there is an entire class of problems whose solution must come primarily through governmental actions; these problems exist both for individual nations and for the international consuming community as a whole. We can increase the available supply of crude oil and of substitute fuels, but to do this would demand capital investment on a vast scale. Such investment constitutes an enormous threshold and is frequently the only barrier between where we are and where we would like to be. It is extremely difficult to envision private investors being willing to create large-scale mass transit networks, develop energy conservation-oriented communities, or construct synthetic fuel plants. Even where the cost of conversion to substitute fuel is not great, poor consuming nations may be unable to make the initial investment needed to reduce their dependency on OPEC oil. We need a mechanism to finance the programs of investment that make it possible to kick the oil habit or reduce it substantially.

One highly important but paradoxical aspect of this need figures prominently in discussions of national security and the availability of oil. Logically, because it is a depletable resource, it can best be conserved by not using it. That is to say, the greater a nation's reliance upon imported oil, the more its own oil is potentially available. It is clear, however, that from whatever point of view, military or civilian, the question is not one of ultimate but rather short-run availability. For purposes of bargaining in the oil market the possession of a readily available supply would be enormously useful to consuming nations. The ideal situation for bargaining purposes and for national security purposes as well would be to have stockpiled oil and fully developed domestic producing and refining capacity that could be turned on or off in accordance with the availability of lower-cost foreign oil. Even if they are unused, the very existence of such facilities would serve to force a reduction in oil price.

Naval oil reserves as well as lands leased to private companies could be developed in this manner by the U.S. government; it is hard to believe that any private firm would be anxious to invest in such a system where, except when it was actually in operation, the benefits would flow not only to the particular investor but to all buyers of oil. On a society-wide basis their return on investment would be lower oil prices from current sources of supply.

NOTES

1. The definitive study of the oil market and its institutions is M. A. Adelman, *The World Petroleum Market*, Johns Hopkins University Press, 1972. Data on price trends for imported crude oil and refined petroleum products in various markets may be found in Neil Jacoby, *Multinational Oil*, Macmillan, 1974, and Joseph Yager and Eleanor Steinberg, *Energy and U.S. Foreign Policy*, Ballinger, 1974.

2. A "take" price of $10.12 per barrel was announced at the

Vienna 1974 meeting of OPEC ministers. See Table 3 below for take price data.

3. Walter J. Levy, "Implications of Exploding World Oil Costs," in Committee on Interior and Insular Affairs, U.S. Senate, *Implications of Recent Organization of Petroleum Exporting Countries (OPEC) Oil Price Increase*, 1974.

4. U.S. Department of Commerce, *Survey of Current Business*, November 1974.

5. Jahangir Amuzegar, "The Oil Story: Facts, Fiction and Fair Play," *Foreign Affairs*, Summer 1973, suggests that this split was 18/82 in 1948, 32/68 in 1952 and 50/50 in 1960.

6. Data on sources and uses of energy taken from Joel Darmstadter, "Energy Consumption: Trends and Patterns," in Sam Schurr, ed., *Energy, Economic Growth and the Environment*, Johns Hopkins University Press, 1972, and James P. Grant, "Energy Shock and the Development Prospect," in Committee on Interior and Insular Affairs, *Implications of Recent OPEC Oil Price Increase, op. cit.*

7. 1974 oil accounted for more than 20% of total world trade. Estimate based on IMF data and Grant, *op. cit.*

8. U.S. Treasury Department.

9. Levy, *op. cit.*

10. *Ibid.*

11. *Economist*, December 7–13, 1974.

12. Levy, *op. cit.*

13. Amuzegar, *op. cit.* More recent sources suggest a figure of $2.50 per barrel for U.S. production.

14. Darmstadter, *op. cit.*

15. M.I.T. Energy Laboratory Policy Study Group, *Energy Self-sufficiency: An Economic Evaluation*, American Enterprise Institute for Public Policy Research, November 1974, and John M. Blair, "The Implementation of Oligopolistic Interdependence. International Oil: A Case Study," paper presented to the Association for Evolutionary Economics, December 1974.

16. A good introduction to the study of serious games is R. Duncan Luce and Howard Raiffa, *Games and Decisions*, Wiley, 1975.

17. For detailed description see Hearings before the Subcommittee on Multinational Corporations of the Committee on Foreign Relations, United States Senate, Ninety-third Congress, First and Second Sessions on Multinational Petroleum Companies and Foreign Policy, January 30, 1974, Parts 4 and 5.

18. James E. Akins, "The Oil Crisis: This Time the Wolf is Here," *Foreign Affairs,* April 1973.

19. Noted by M. A. Adelman in "Is the Oil Shortage Real? Oil Companies as OPEC Tax Collectors," *Foreign Policy,* Winter 1972–73, p. 104.

20. Blair, *op. cit.*

21. For an account of the various joint ventures and other links between oil companies, see James Ridgeway, *The Last Play: The Struggle to Monopolize the World's Energy Resources,* New American Library, 1974.

22. Blair, *op. cit.*

23. M.I.T. Energy Laboratory Policy Study Group, *op. cit.*

24. See "Soviet Energy: An Internal Assessment," by Marianna P. Slocum in *Technology Review,* Vol. 77, Oct./Nov. 1974, pp. 16–34.

25. Saudi Arabian Oil Minister Ahmed Zaki Yamani, quoted in *San Francisco Chronicle,* December 17, 1974, p. 12.

26. See testimony of John McCloy, Senate Hearings on Multinational Corporations, *op. cit.,* Part 5, p. 65.

27. For discussion of pertinent U.S. tax provisions see Senate Hearings on Multinational Corporations, *op. cit.,* Part 4.

28. See David Howard Davis, *Energy Politics,* St. Martin's Press, 1974.

29. For an insight into the extent of company participation in the embargo see "Cutoff of Petroleum Products to U.S. Military Forces," U.S. Senate, Committee on Government Operations, Permanent Subcommittee on Investigations, Part 8, April 22, 1974.

30. Peter R. Odell, *Oil and World Power,* Penguin Books, 1970.

31. Adelman, "Is the Oil Shortage Real?" *op. cit.*

32. Testimony of John McCloy, Senate Hearings on Multinational Corporations, *op. cit.,* Part 5, p. 273.

33. Yuan-Li Wu, *Economic Welfare,* Prentice-Hall Economic Series, 1952, Chapter 6.

34. Adelman, "Is the Oil Shortage Real?" *op. cit.*

35. Amuzegar, *op. cit.,* p. 679.

36. Odell, *op. cit.,* and Adelman, "Is the Oil Shortage Real?" *op. cit.*

37. Adelman, *ibid.,* p. 106.

38. Amuzegar, *op. cit.,* p. 683.

39. John Maynard Keynes, *The Economic Consequences of the Peace,* Macmillan, 1972, pp. 9–13.

40. Adelman, "Is the Oil Shortage Real?" *op. cit.,* and Yamani, *op. cit.*

41. National Energy Project, *Dialogue on World Oil,* American Enterprise Institute for Public Policy Research, 1974, p. 25.

42. Hollis Chenery, "Restructuring the World's Economy," *Foreign Affairs,* January 1973, p. 242–263.

43. Adelman, "Is the Oil Shortage Real?" *op. cit.*

CHAPTER 4

WHAT CAN BE DONE?

WE FACE a world in which all but a few oil-producing countries may be poorer than they need be or would have been—among some of the poor, mass starvation, among the rest, instability. Populations which cannot understand sudden deprivation may see the rise of class and racial conflict. If governments cannot cope, they are likely to change rapidly as they find it impossible to manage either domestic conflict or international finance, vacillating between adventurism and appeasement. They may welcome oil money only to be repelled by visions of foreign tutelage. They may seek aid only to be told they are not good credit risks. If they repel aid they probably will discover they lack foreign exchange to buy what they need. Recognition of the plight of still poorer nations may come up against the desperate need to maintain living standards at home. The richer the OPEC nations become, the more they stand to lose; as their worries mount, they seek greater armaments. They too are caught between the urge to exercise their new powers and the fear that they have become the world's most promising targets. As its opportunities grow, the Soviet Union is likely to increase its commitments to the Arab nations in OPEC. The initially small Soviet stake in the Middle East may become progressively larger—first to hold on to advantages that have become huge, and then to extend them by sharing the wealth of those it has protected. Yet, can it control the antipathy of its protegés toward Israel, their understandable desire to make their power equal to their finances, or their anxieties? Any or all of these may tempt these nations to threaten early, attack first, or nervously run with their money. These postures could force instabilities on the rest of the world which would set the stage for war. War may come, not from calculation but from inadvertence.

No one has learned to adjust to the new circumstances. Lines beyond which others must not go have become blurred. Alliances change too rapidly for friend and foe to become established as such. Miscalculations are made. The United States sees an opportunity to recoup its losses, the Soviet Union to hold on to old gains and to make richly rewarding new ones. Here is a recipe for disaster: mix well, keep shaking, and pour on the oil. With atomic reactors scattered hither and yon, some given in tribute, others for cash; with Israel and OPEC fearing their futures; with the United States alternating between anger and despair, atomic war could become substance rather than remain shadow.

What can the United States do to arrest or reverse this systemic decline? What should it do? What would be *right*? Action depends not only on possibility and feasibility but also on morality. The nation's leaders and its people must believe it is right as well as prudent for them to act.

Today the morality of the oil producers is taken for

What Can Be Done?

granted. No one questions their right to raise their prices eleven and more times, to limit production, even to cut it off altogether. It is their oil, apparently, and they can do what they want with it. The oil producers argue, in addition, that previously they were underpaid for their product, that they are only compensating for inflation in the commodities they buy, and that, like any commercial operation, they are entitled to charge as much as people are willing to pay or the traffic will bear. OPEC oil ministers are fond of reading Western oil ministers introductory lectures in economics. Perhaps the standard texts need revision.

Whatever may be said for the oil producers, they certainly are not engaged in an ordinary commercial transaction. Nor did they get into the big money in peacetime. They raised their prices in the midst of the October war as part of an embargo deliberately designed to bring pressure against Israel. One can talk forever about what OPEC might have been able to do, leaving out the war, but the historical fact is that they acted as part of war. Indeed, the oil embargo may have been the most decisive part of that war—the part that led to an unfavorable diplomatic outcome for Israel, and that continues to have the most far-reaching consequences.

To say that the price rise was part of a war is not necessarily to justify a new war to undo what was done in the old. What was done in war, to paraphrase Secretary Kissinger, may be undone in war, but that does not mean it *ought* to be. All that has been shown so far is that there is no merit in saying the oil price rise ought to be accepted because it is part of the usual commercial rules of the game to which the oil importing nations had heretofore subscribed.

The *ex post facto* rationalization given by oil producers for their price increase is that it corresponds roughly to the rate of inflation since 1945. Oil prices, they say, have remained stable, but the goods bought with oil revenue have gone way up. This is one way of looking at it but not the only, nor even the usual, way; that would be to compare the cost of extraction per barrel with the payments received by the producing countries. It costs so little to get oil out of the ground in the Middle East (10¢ a barrel) that the pre-increase return (seven to nine times the cost of extraction) was already staggering. The real justification is that oil producers can get together and consumers (thus far) cannot. But then we leave the moral realm and return to *Realpolitik*.

The inflation argument does not stand up even on its own grounds. Doubling, not quadrupling, oil prices would more than make up for any inflationary effects OPEC countries may have suffered. As B. Nagorski puts it:

> There is an enormous difference, a difference in scale, in the quantities of commodities involved and in financial results. . . . Let us assume that OPEC members are importing five million tons of grain in a given year, while the price of grain increased from about two dollars per bushel to five dollars, or in rounded-up figures, from $80 per metric ton to $200. The gain importers would need $600 million to cover the price increase.
>
> Yet the combined production of all OPEC countries is in the range of thirty million barrels per day. One single dollar per barrel would yield a revenue of $30 million

What Can Be Done?

per day or almost $11 billion per year, eighteen times more than the expenditure for higher costs of grain. To put it otherwise, 5½ cents per barrel would provide the $600 million required for this particular purpose.

The price of cement, another basic commodity, rose from about $30 per ton f.i.o. Persian Gulf to approximately $70. Up to five million tons of cement may be imported yearly by the OPEC countries. If so, $200 million would be needed to cover the $40 per ton difference in price, a sum roughly equivalent to the yearly revenue from two cents per barrel of oil.

. . . Higher prices of industrial equipment and consumer goods imported from advanced nations can by no means be considered as a valid justification of a price increase of oil. . . . A few cents per barrel would compensate the OPEC countries for higher costs of each major commodity, while less than one dollar per barrel would probably provide enough revenue for covering the total cost increase of all imports.[1]

Behind the mild request for equity, moreover, lies a fierce demand for superiority. The gigantic gains of the recent past are to be guaranteed into the distant future. Alone among the nations, oil exporters are to be insured against calamity. The feast OPEC has prepared for itself includes demands for guarantees against expropriation (without which it could not have enriched itself) and against inflation, to which it has become the main contributor by its high prices. OPEC is allowed to fuel inflation but not to be consumed by its own fire. That all nations must suffer the consequences of their own actions is a reasonable moral principle, but OPEC is to be excepted—and it is an unreasonable exception. If OPEC is to be protected against expropriation of investments, so should all other nations; thus no one else will be able to enrich himself by doing as OPEC did. That hardly seems fair. If other nations insist on following OPEC's example, by confiscation whenever convenient, OPEC will not be able to invest with safety. This in turn will lead to further disruption of the world economy, as OPEC tries to move money just ahead of the expropriator. That hardly seems right.

But, OPEC claims, the poor are entitled to become rich if they can. Nobody denies that. The question is whether OPEC's actions are in the interests of the world's poor or, more generally, whether most nations and most people will be better off (even if a few nations and people are worse off) because of what OPEC has done. Now OPEC is not exactly a Robin Hood, taking from rich nations to give to poor. Some members of OPEC, such as Abu Dhabi and Kuwait, are rich by any standard. Indeed, they look more like banks built above lakes of oil than countries. Others, like Saudi Arabia, Iran, and Libya, are far above the average income per capita of the world's peoples. And the price OPEC charges is the same for rich and poor alike—reminding one of the old story in which J. Pierpont Morgan and a homeless man are equal because both lack the right to sleep on a park bench.

OPEC negates the doctrine of comparative advantage under which nations produce the things at which they are best, and trade for the rest. It does not pay a nation to produce at home, at four times the cost, goods it could import at one quarter the price; yet this is precisely what will happen as OPEC forces nations to realize they cannot go on depending on others for essential

What Can Be Done?

commodities lest these be expropriated, embargoed, or supplied only at ruinous prices. Who will win in this Hobbesian trade war of all against all? A few poor, perhaps with essential raw materials; the vast majority will not since they either lack what others need, or will be less able to compete. Forewarned is forearmed. The rich will be ready; the poor will be ruined.

The one certainty is that the world as a whole will be poorer with higher prices for energy than it would otherwise have been. The short-term effect of the increase in the price of oil will be to make poorer all but the oil exporters and a few other countries that are self-sufficient. Everyone else will suffer a decline in standard of living equal to at least $100 billion a year in additional costs. Because much of the money will go into armaments or be kept on short-term deposit, the overall loss to world production will be even greater. And the long-term loss will be even greater still. Every nation that can do so is now scrambling to provide alternative sources of energy that are financially feasible at the equivalent of $11 plus per barrel of oil. Before long, these countries, including the United States, will be the strongest supporters of oil at that price so as not to undermine their vast subsidies for alternative sources. Thus the world will saddle itself with permanently high energy prices. After a decade or so, poor countries will not be able to become richer by exploiting oil resources. In the midst of an energy glut, rich countries will refuse to import oil so as to use up their own domestic supplies of energy. Who then will suffer most from high energy prices—developing countries striving to make their way up, or developed ones guarding against going down? The answer is as depressing as it is obvious.

OPEC's reply to the charge that it is further impoverishing the poor is self-preservation: the oil supplies of its member nations will become depleted, possibly within the next thirty years, and they are entitled to make the most of it while they can. The argument shifts from the positive benefits for large parts of the world to the negative costs imposed on oil exporters. Natural law gives way to the law of self-preservation. OPEC is right for itself but wrong for the world. And that is pretty much the way things are. Without further examination, however, it would be unwise to accept the claim of rapid depletion at face value.

Let us suppose the price of oil had not suddenly increased by a huge amount. What would have happened to the demand, supply, and price of oil? Demand was growing rapidly. So was supply, but not as quickly. The market would have responded by taking the price higher. Exploration—not at the frantic pace of today, but at a quick pace nonetheless—would have been encouraged. Profits at these prices would still have been higher than in almost any other industry. More oil would be discovered; this is happening now just as it has happened before, to belie all those previous predictions that the world would be running out of oil, predictions from the 1920s on. Neither the proliferation of expensive arms—which rich countries sell to maintain their income—nor of nuclear reactors (to provide domestic supplies of energy) would have occurred. Alternative energy sources would have been brought in

What Can Be Done?

slowly as technology developed and cost was made feasible. Mechanisms for controlling arms and supplying energy would have had a longer time and a better opportunity to work. For all anybody knows, the OPEC nations would have discovered more oil and been better able to arrange more orderly development over a longer time span for themselves. But there's the rub. They might have done as well for themselves, but then again they might not. Certainly they could not have become so rich so fast any other way. OPEC may be opportunistic, but why shouldn't it take opportunities where it can find them?

The OPEC position is that the oil belongs to them; it is pretty much all they have, and they have a right to do whatever they want with it. To argue otherwise appears to deny sovereignty. And the right to control one's own natural resources strikes such a responsive chord among other Third World countries that they have supported the position despite the fact that they themselves stand to suffer from its immediate implementation. In fact, all countries defend the idea of sovereignty, the idea on which the nation state has been built. To speak of sovereignty, however, is not to end discussion of oil delivery and pricing but merely to begin it.

General principles do not necessarily answer for particular circumstances. Sovereignty is a fact but so are international agreements. Unless nations are to live entirely self-contained lives, in which event oil in OPEC countries would be close to worthless, they must arrange to exchange goods and services. This exchange is regulated by agreements that each party is obligated to keep. Of course, the stronger parties may violate them and the weaker may have to suffer. Here, however, we are speaking of right not might. OPEC collectively and its member nations individually entered into long-standing agreements to supply oil. They broke these agreements by repeated expropriations, price increases, and refusal to deliver. If OPEC invokes sovereignty, its trading partners by the same token can invoke contract. Two rights do not make a wrong: that can be done only by unilateral action which imposes one right over another.

To concede that the ultimate right to dispose of national resources belongs to the country in which they are found does not dispose of the contractual obligations into which it has entered. Had oil exporters notified the then-existing consumers of oil that they would not renew existing contracts, the companies and countries involved could have made alternate arrangements to cushion the impact. Presumably this was precisely why no advance notice was given. Do not nations that possess resources which have been discovered, developed, operated, and owned by foreigners have a moral, as well as legal, obligation to allow these resources to be used for a reasonable period after notice? If not, anybody can seize anything and no international investments, including those of the oil producers, would be safe. Offering some sort of financial compensation is no answer because it covers at best the initial investment—not the present real worth—reimbursing the stockholder but not the citizen. The loss of capital may be compensated, but economic and social collapse cannot.

What Can Be Done?

Imagine that the Soviet Union had gone abroad to buy land with which to alleviate the hunger of its people. Imagine also that it had after many years come to depend on the produce. Would the Union of Soviet Socialist Republics allow the nations within whose domains these lands were found suddenly to withhold their fruits or to pyramid the price—straining its ability to pay, worsening inflation, increasing unemployment, and mutiplying hunger? We think not. Are agreements to be kept only with those confident enough to demand it and strong enough to enforce it?

OPEC might well deny the whole thing. The so-called agreements were forced upon them, they might say, when they were unable to resist. The truth of this may be doubted or affirmed—it makes little difference to the present inquiry. For if oil had not been discovered and developed where it was, it would have happened elsewhere. If no other places had existed, oil today would be less used and coal or some other source of energy would have taken up the slack. Would the OPEC nations be better off now if their oil were still underground and demand were so low as to justify keeping it there? No one doubts they have benefited from the development of their resource.

All OPEC has shown, really, is how exploited member countries have traveled in rags on the road to riches. One could just as well make up a story in which OPEC enticed the world into overdependence on oil through low prices, then withdrew the sticky substance so that the addiction they helped create could be managed only by much more money.

OPEC's actions hurt the world's poor people and its

democratic governments. No decent international order would permit huge income transfers from the poor and democratic to the rich and autocratic.

The oil belongs to OPEC's governments only by accident. They did not discover it, drill for it, pump it up, or pipe it out. They broke treaties, contracts, and understandings to enrich themselves. Their Arab members used oil as a weapon of war. They do not deserve to be rich; they deserve counteraction to drive their oil prices down.

Acting against OPEC is no gurantee of virtue. Actions do not become good by being directed against evil. Neither should evil be suffered because of pretension to good. No nation adversely affected by the oil exporters need fear that it would be behaving badly in taking countermeasures, whether these be economic, to bring down the price, or military, to assure supply, or any combination thereof, to prevent systemic decline.

But would not use of force for commercial considerations be immoral in and of itself? Life, according to some arguments, may be taken to protect life but blood may not be shed to preserve property. By this line of reasoning, it would be immoral to make war in order to make money. Economic reason can never justify invading other countries; the fact that war in the past originated in commerce can only lead us to hope that mankind has (or should) become more civilized. Men might fight to protect their basic liberties but not, or so it appears, to save their holdings.

The great irony is that only an affluent society can afford to devalue the importance of property in the way

What Can Be Done?

that such an argument does. For if property were believed essential to that "Life, Liberty and the Pursuit of Happiness" the Declaration of Independence describes as the rationale for instituting government among men, no one would doubt that it was worth defending.

Property is not just "goods"—it is a form of power. But whether it is used for plunder or for protection, neither life, nor liberty, nor happiness is possible without property. There was a time when this was understood by everyone; it is worth bringing up now only because it is so rarely mentioned. When one thinks of property not merely as land and buildings but as a call on assets, it is clear that "owning" and "living" are two ways of saying the same thing. Without the freedom to dispose of one's assets, one has no liberty. Just as the ability to switch votes from one candidate to another is a cornerstone of political rights, just as the right to refuse employment is integral to economic freedom, so property is essential to personal liberty. What difference is there between forcing a person to work for you and compelling him to turn over his wages? What difference does it make if food is forcefullly taken from a person or he is deprived of the property by means of which he would acquire food? Would it matter if property were purloined from a peasant or captured from a collective? The results—loss of life, denial of liberty, inability to seek fulfillment—would be the same. To say that our property is not worth defending is to assert that we (our lives, our liberties, our pursuits) are not worth protecting.

But, when you come down to it, about whose property, whose fortune, are we speaking? We are so mesmerized by large institutions—governments, oil companies, energy-producing and delivering utilities—that we fail to see the trees for the forest. It is easy for an OPEC statesman to point, with ironic glee, at the "sufferings" of "rich nations," "rich companies," and "rich utility-monopolies." But institutions themselves do not suffer; the suffering is done by persons. Those who are now suddenly and traumatically afflicted, in their property and fortunes, are individual human beings. It is possible even for "rich" nations to have poor people; if they are today suddenly much poorer, it is in some large measure due to what we are describing.

Nevertheless, as the richest large nation in the world, the United States is hardly in a position to cry poverty and command a sympathetic world audience. If America were to "poor mouth," few would listen. Were it to act to enrich itself at the expense of others, it could hardly dignify this behavior in universally valid moral terms. Is there, then, no interest, short of an attempt at conquest of its heartland by some foreign foe, that the United States can cite in its self-defense?

The historical answer, since the Second World War, has been a sublimated self-interest: the Government of the United States responds to calls of distress, which may be in its interest to answer, but does not aggressively affirm its own interest by independent action. From Korea to Lebanon to Vietnam, the rationale (or the pretext) has been an invitation to come to the rescue of another government. Now, whatever game is played in the Middle East, an invitation to invade is not in the

What Can Be Done?

cards. Unless Israel intervened, the United States would have to act entirely on its own. Instead of covering its activities with a foreign cloak, the United States would have to assert its naked interest. The Emperor may never in fact have had any clothes, but his own belief that he did afforded him the kind of assurance that has been noticeably lacking in recent times.

Enemies are expected to oppose, allies to support, their alliance: entry without invitation from OPEC is understandable; no-entry from NATO is harder to take. The United States has a general interest in having allies, and a specific one in being allied to viable democracies. If Western European governments retained their independence but lost their democratic character, the United States might maintain its military might but would lose a supportive environment of free societies. Since they are more greatly afflicted by the oil crisis, America's allies take a different view of their self-interest; however, the United States now is in the position of acting for them without their consent. Presuming to act in the interests (but against the express preferences) of allies is another reason for American hesitation. For if the United States is to act at all, its actions must be of, by, and for itself.

As of this writing, the seriousness of the new international crisis has not yet prompted any major figure in American public life to publicly address the subject in the candor, comprehensiveness, and gravity which it deserves. Not long ago at Fulton, Missouri, Senator William Fulbright delivered some valedictory remarks of this sort at the end of his long congressional career.

The Great Détente Disaster

His diagnosis of the nature and causes of our current predicament, and his prescriptions, differ in many particulars from ours but his general assessment does not.

> If adversity were all it took to get people to behave magnificently, or even sensibly, America and the West would be at this moment within minutes of their "finest hour." That, however, seems not to be the case, as we confront a different kind of adversity. In addition to his own courage and eloquence, Churchill had the added assist, in Hitler, of all that one could ask in the way of a villain.
>
> Today we face a different kind of adversary, not a tyrant but a condition and the ominous prospects to which it gives rise. The condition is economic crisis, and the prospects—unless immediate and drastic corrective measures are taken—are for economic depression, political breakdown, and perhaps war.[2]

What is also important to note is how little attention was given to Fulbright's startling comments when they were uttered. Daily attention is paid, in our press and in government pronouncements, to particular segments of the crisis—notably to the means by which the American nation now can cope with its own energy needs. The acute interest given this aspect of the problem is well deserved; it attests to a revived sense of American pragmatism: having defined a "problem," it is possible then to look for its "solution." But also—*this* problem, and its possible solutions, is one which political figures can handle safely without too much risk to their reputations. Diagnoses and prescriptions may be made which appeal to those Americans—and there are increasing numbers of them—who now see America's role in the world more and more in protectionist or

What Can Be Done?

isolationist-nationalist terms. First painted by Nixon in early 1974, the vision of an America self-sufficient in energy is reassuring to everyone, with the possible exception of those on the outer fringes of the ecology movement for whom aesthetic aspects of nature are more to be cherished than are the basic requirements of tolerable human existence. Most of us would sleep better knowing that ten years from now America may be what it is not today: independent of the outside world for its most essential raw-material requirements. This kind of hope, however, is one which North America shares with no other advanced industrial region outside the Communist world. For all the others, the prospect of self-sufficiency is out of the question.[3]

Another reason for the attractiveness of the quest for American self-sufficiency is that, constructively absorbed in it, one may thereby successfully avoid facing more unpleasant ramifications of the world crisis. If America in the Depression years (as Will Rogers said) was the only nation in the world in which one could drive to the poorhouse in an automobile, so America in a fractured world of neomercantilism, trade wars, and conflicts could be the one nation whose citizens might retire to sumptuous air-conditioned hurricane cellars.

In wide sectors of the American public, the chief solace of the Nixon foreign-policy years was the prospect of a dawning era of negotiation, not confrontation. In his direst hours, just before resigning, the president took what little comfort he could from the approbation he had received for tuning down the engines of American foreign policy to a level compatible with the wishes of a divided and disenchanted public. Whatever price may have been paid for this in the international scene, certainly the calculation was well chosen in terms of its domestic reception. If the polls are correct, the president's great gamble of overtures to Communist China was welcomed even in the most conservative circles.

Our political opinions, while they arise as specific responses to concrete experiences, events, and choices, quickly jell themselves into a generalized body of sentiments. Sentiments formed in consequence of recent experiences are those most likely to govern our ways of evaluating new conditions. They provide stereotypes of ourselves, our moral qualities, and of our estimate of what we can accomplish. Often when traumatic national events have prompted such soul-searching, the dogmas governing our future behavior are found frozen into negative, rather than positive, axioms. Was it a strong presidency which brought America into World War I, or into the Vietnamese war? Then strong presidencies are out. Was it a series of small events, leading from slight commitments to giant ones, which led to the vast involvement of U.S. military forces in Vietnam? Then no such small steps should be undertaken in the future. Was it that the great, futile commitments of U.S. will and energies in Vietnam simply were a manifestation of the swollen "universalism" of American commitments—an "undifferentiated globalism"? Then America's systemic commitment of wide proportions itself must be dissolved. "Come Home, America" was the slogan of the 1972 McGovern campaign. But it echoed sentiments far broader than those of his adherents. "Full of much of what not to

What Can Be Done?

do," we are still full of the negative lessons of Vietnam. But, as Mark Twain once cautioned:

> We should be careful to get out of an experience only the wisdom that is in it—and stop there; lest we be like the cat that sits down on a hot stove lid. She will never sit down on a hot stove lid again—and that is well; but also she will never sit down on a cold one any more.

In truth, there is no crisis worse than one in which it is widely supposed that the will to meet it is missing. In political matters, as our tutor the Shah of Iran has pointedly remarked, Vietnam has shown the American public the lesson of *all* interventions: they are futile.

It is the nature of political will that it consists not in momentary strategies but rather in fixity of purpose toward some knowable end. Will cannot be said to exist, or to have any particular value, if it is unattached to objects, and is not inspired by a clear devotion to their attainment. So also with respect to power: power lacks meaning when unrelated to plausible ends. Otherwise it is indiscriminate and unfocused energy, easily dissipated. Will thus is related intimately to power, and power to knowledge. But practical knowledge cannot consist solely in negative injunction. In the world of action it must consist in some positive conceptions of what is to be attained, what preserved, and what to be deterred or deflected.

In the traditions of American foreign policy, many writers and statesmen have pointed out the all-too-frequent disjunction in public opinion between aspiration and actuality—ideals and reality—between asserted goals and the means and costs of attaining them.

The Great Détente Disaster

The inspirations of Jefferson and Wilson grate against the cautions and warnings of Hamilton, Theodore Roosevelt, and Acheson. Often the disjunction can become intolerable. Yet it might be generally agreed that neither ideals nor self-interest and cautionary warnings alone suffice to sustain national action in affairs of the world. The risk entailed in inspired idealism is that it may be blind to the costs and likely consequences of action, and lead to profound subsequent disillusionment. The risk of prudent self-interest is that its pursuit is hobbled by absence of enthusiasm; it suffices best in instances where threats to survival interests elicit a profound, self-evident, and widely shared fear. A realistic, widespread, and self-evident fear may, on occasion, prompt an extraordinary popular consensus and even a great and unitary willingness to sacrifice—as was true for all classes of Englishmen in the Battle of Britain. Heroism is not a virtue for normal times; it emerges and enlarges only as desperation inspires and demands it. But the inspiration which despair often generates is not so precious a quality that one deliberately seeks or welcomes the harsh circumstances which elicit it.

These abstract remarks are by way of suggesting that will is possible only when knowledge is present to suggest the nature of purposes; it is a will-to; and if much of our current thinking about world politics is still conditioned by now-obsolete recent experiences, by a will-not, it is dominated by the immediate past rather than focused upon the future. Our chief difficulty, furthermore, in obtaining a view of what is now at stake is that past experiences in no way provide significant clues to what now is in store for us. If our poli-

What Can Be Done?

ticians, then, are silent on these matters, it is perhaps because many of them, too, are genuinely confused.

For in point of fact, to think about the unthinkable is different now from the thinking for which Herman Kahn once argued: we are compelled now to face matters of far greater complexity than the simple, stark, and terrifying scenarios of nuclear bipolar war which Kahn, Cassandra-like, painted in the 1950s.[4]

There are other reasons for the fact that the OPEC-engendered crisis has thus far been so passively accepted. The energy crisis has come on the American public at a time when crucial sectors of Western opinion elites have been mesmerized by the not unimportant question of the "limits of growth." In the context of this book it is impossible to deal adequately with the vast ramifications of this controversial issue. But the most troublesome aspect of it turns on the question of whether economic growth—conceived in terms of material production, consumption, and innovation—is compatible with ecological safety. It turns also on the critical question of *general* impending resource scarcity. The expectation of continuous growth and confidence in its results have been key characteristics of our civilization since the mid-eighteenth century. Now we are urged to repudiate them. In the recent words of Arnold Toynbee:

> If mankind is to salvage itself, it must try to return to the state of relative innocence in which it was living before the outbreak of the Eighteenth Century ethical and technological revolution in the West. In order to execute this difficult manoeuver, it will have to look for some inspiration, guidance, and example either in the pre-indus-

trial West or in some non-Western region in which the pre-industrial way of life is still a going concern.[5]

A certainty of impending doom, inspired by some widely read theoreticians such as Robert Heilbroner, has led a number of people to argue the merits of planned scarcity; rigid, even authoritarian means are suggested to check the appetites of gluttonous consumers. The implication of the called-for economy of managed scarcity and constraint suggests far more than the control of appetite; it suggests also the control of all human (Promethean?) qualities of invention, innovation, and laborious construction—for if these qualities are permitted to persist without check, so the argument proceeds, they may lead to a devastated planet.

It goes without saying that this argument does not lack opponents; it is also a special First World argument. It does not appeal to the leadership of nations continually plagued by threats of famine. It does not attract the billions whose livelihood depends upon a dynamic economy. While it has resonance in important quarters of the advanced societies, guiltily aware of their own comforts, it has none for the millions of jobholders who might be immediate victims or to the many more millions in the poor countries dependent on them for livelihood.

But certainly the gigantic OPEC-induced crisis, when viewed through ecological lenses, seems not wholly a disaster. The actions of the new oil monopolists can be made to appear as those of a benevolent *deus ex machina,* visited upon us in order to command us to do what in any event the logic of no-growth tells

What Can Be Done?

us voluntarily to do. It is thus a timely figure to enter the stage of our national life; the severity of its demands could inspire a response of austerity.

Some of us also, if to a lesser extent, have been mesmerized by another issue—that of "historical equity." The oil-exporting countries of OPEC have all, in some way or another, been the objects of Western exploitation in the past. They have long been combed over by Western interlopers—prospectors, developers, business firms—and, in most instances, affected by Western political control. In this sense they all can be regarded as historical *subjects*. History was not on their side before; now it is! Why should not a long season of subjection entitle them to a new season of rectification? A Western sense of "imperial guilt" can in this regard also be locked together with a diffuse sentiment, in most of the Third World, which delights in this situation of reversed fortunes. (Why, for instance, does the government of Bangladesh—a nation now grievously victimized by the oil price hike—continue to lend public support to its Arab instigators, if not for this very reason: the sensed need for solidarity against the West, based upon past grievances, is even stronger than Moslem brotherhood.) This argument is not without supporters in the West; it anesthetizes our wounds; it paralyzes our ability to think of any measures which we might take, should such measures threaten to offend this diffuse *ressentiment*. We are truly the victims of our past, or at least of a past which we have been taught to think of as filled only with our misdeeds.

The chief problem with such sentiments—the hope for ecological rescue on the one hand and the dream of equity on the other—is that the former speaks of supposed far-distant future calamities, while the latter addresses itself to past misdeeds. Together they serve to inhibit our ability to deal with matters now at hand. It would serve no nation well were our ecological difficulties "solved" by the collapse of the world economy. A few primary-product producers might reap short-range gains from initiating giant price rip-offs; and some psychological benefit might come to now-disadvantaged nations, from the spectacle of widespread suffering and reverse humiliation in affluent Western societies. Some Western intellectuals might also see in such a spectacle the judgment of history. But such benefits—if benefits they are—would surely be of short duration, especially if the chief consequences of these actions were the systemic ones which we have outlined.

In any case, large swings of mood and attitude have long dominated our national life. In the past they have oscillated between intervention and introversion, between support for activist international programs and opposition to them. Some scholars have gone so far as to discern a cyclic quality to these shifts, suggesting the possibility that there is at work a kind of generational reversal of magnetic fields. But such an interpretation savors more of astrology than of science. American moods about foreign policy are not autonomous; as manifestations of a literate and democratic society they are highly responsive to events, and the crucial events are those which occur in the international environment. Swings of opinion may be too exaggerated—between inactivism and punitiveness, between cynicism and idealist messianism; but they do not arise

What Can Be Done?

autonomously, like the thought patterns of a manic depressive. They are likely to continue as long as America remains an open and democratic society. And this is because in any crisis situation invariably there are choices, and the literate public knows it.

This being true, it is fruitless for present purposes to cavil about a timeless characteristic of opinion, and more to the point to summarize the situation of public opinion right now. A recent article in *Foreign Policy* by William Schneider of Harvard University suggests that our current climate of public opinion may differ sharply from that of nearly every previous time of crisis except the American Civil War.

If Mr. Schneider's analysis of poll data is correct, what distinguishes current public moods about foreign policy from previous ones is the fact that—until Vietnam—"debates over foreign policy at the elite level have usually not . . . resulted in the mobilization of mass political forces in such a way as to enlarge the conflict." Furthermore, the emergence and persistence of antimilitarism as an attitude represents "a basic change in the traditional structure of foreign policy opinion." The antimilitarist trend has not surprisingly been accompanied by a strongly isolationist trend.[6] Another feature of our current situation is a largely unconscious but effective collusion between a new antimilitarist liberalism and a growing sentiment among the poor and poorly educated of "mistrust of leaders on such issues as aid, and hostility toward all international involvement." More than the rich, the poor abhor war. The combination of these tendencies, Schneider suggests, makes possible the development of a profoundly strong populistic nationalism. "An atavistic 'America First' economic movement is entirely consistent with the isolationist mood of the mass public." "History," he concludes, "provides many examples of economic discontent translated into foreign policy crusades, especially when no workable domestic solution is forthcoming. (The oil cartel is the obvious target, and not an altogether irrational one.)" [7]

But how could the less formally educated and the comparatively more poor—who do not like the wars they are asked to fight—be brought to support aggressive action? An apparent empirical inconsistency provides a clue: many of the same people who favored withdrawal of all American forces from Vietnam also favored heavy bombing and mining of harbors in an effort to bring peace. Schneider reasons that "forceful solutions are likely to be acceptable to lower-status 'isolationists' if such measures are taken for the purpose of ending international involvements, rather than extending them." [8] Armed force in defense of American national interests in maintaining oil supplies at affordable prices for the domestic economy to fight inflation and battle unemployment might well receive substantial popular support; a long-term occupation of alien land would not.

There are more than considerable dangers in this trend toward what might be called "iso-populism." For one thing, it could portend a permanent collapse of traditional American liberal views of world politics. Patriotic Americanism could be married to a form of constructive internationalism when both the former and the latter were heavily influenced by a Jeffersonian con-

What Can Be Done?

fidence in the compatibility of long-term historical change in the international community with long-term historical progress within the United States. But were this connection to be broken, advocacy of both economic and political internationalism, long a hallmark of American foreign policy, could easily give way to a harsh, inward-looking Republic, which rejected the assumptions of international order asserted by two generations of American statesmanship. A short-lived American disillusionment with its Vietnam experience might then give way to a long-term American aggrieved hostility to "malign" external forces. It might be regarded as something of a misfortune if the break-up of the world economy necessarily had to be laid at the door of certain Third World countries, but it would be an unmitigated catastrophe if this gave rise to a view of things which saw America and the Third World as a whole as undifferentiated adversaries. An America stripped completely of its Jeffersonian convictions, yet remaining the most powerful nation in the world, would not be a happy or comfortable member of the family of nations. It might be more feared and respected than it is now; but certainly it would help make for a far more troublesome (and a less just) world.

What, then, as is suggested by the title of this chapter, is to be done? The answer cannot be arrived at without a serious and sober consideration of the use of force.

Many of us place peace above all other virtues, some of us even above our own interests. In times of really severe duress, however, such sentiments are drowned out—especially if they become clearly irrelevant to circumstances.

There are others among us, to be sure, who are exhilarated by warfare, either by the experience of it or in their private fantasies. Some are even more grandiose: to them, mortal combat is the crucible in which mankind refines noble instincts of courage and self-sacrifice.

Precisely because there are such people among us, decisions about whether or not to use force, and how and when to do so, must be made by men who gain no slightest satisfaction or enjoyment from violence as such. Fire fighting is not a profession for pyromaniacs; nevertheless, firemen on occasion do use fire to create firebreaks.

So on occasion force must be threatened or employed when far more unfavorable and forceful tendencies need to be checked. The damage wrought by force may be less than the damage resulting from a dogged determination to avoid its use.

The worst thing the United States conceivably could do, in current circumstances, for itself and the world would be to make military moves for which it was politically unprepared. This is ordinary prudence. Prudence should not take the form of vacillation, but must carefully weigh likely consequences of various available courses of action, including inaction.

Certainly, no one abroad—allies or adversaries—will believe the United States to be serious about any course of action until its government has established the economic sacrifices necessary to show that it is serious about oil. At home, no one could responsibly entertain the possibility of using force unless he were convinced of the futility of other, less risky, or more attractive measures. Developments outside America's control, however, might compel actions on its part which few even now find acceptable. A renewed Middle Eastern war, with a second-round embargo, might be such an occasion. Then, were aggressive OPEC attempts made, for instance, to choke the outward flow of Iranian oil from the Persian Gulf, the shoe would be on the other foot. But that is only one of many such possibilities.

Half-hearted demonstrations of force, or the half-hearted use of it, could well be worse than any other course of action. It would be bound to fail. The ensuing lapse into passivity would be humiliating. Even worse, however, would be an occasion in which the United States, acting with *force majeure,* attained short-range battlefield objectives only to find in victory that the whole world had turned against it, and that it was alone (perhaps with Israel) in a hostile political environment.

These two dire possibilities, flowing from the use of force, seem more worthy of our serious attention than another which is prominently mentioned: that preemptive American actions on behalf of oil would trigger a direct U.S.-Soviet confrontation. For the area of which we now speak is not, and never has been, one in which Soviet vital interests have been invested. In fact, the regimes of the area long have defensively resisted Russian encroachments. That the Soviet Union, like Tsarist Russia, traditionally has aspired to hold sway over these areas certainly has not meant that the continued frustration of that goal has in any way undermined the security, or even the economic integrity, of the Russian system. In the recent past the United States has accepted Soviet definitions of threats to its security within its proclaimed sphere of interest. If America's and the West's vital interests now are threatened in the Middle East, there would be ways of conveying that knowledge to the Russians. Indeed, the basic understanding between the two superpowers has long tacitly but firmly entailed a respect for questions pertaining to each other's most basic necessities in their respective spheres.[9] That the Soviet Union might expect to reap political advantage from a Middle East war must be expected. But that is hardly the same as the prospect of direct Soviet intervention, and the threat which that would pose of a superpower confrontation. (It needs hardly to be mentioned here that the Russians do not depend on OPEC's energy for their livelihood. Even now the Soviets continue to sell their own oil to European neighbors at rates far below those imposed by OPEC on non-Communist states. Future Soviet-U.S. collaboration in energy-resource development, perhaps entailing significant U.S. capital transfers, might ease whatever concerns Soviet leaders would come to have, that at some distant time the frustration of their Persian Gulf aspirations would foreclose their access to needed energy resources.)

But of course we know that the difficulties are far more complex than the "mere" issue of oil and power politics. One procedural problem which has afflicted U.S. Middle East policy since the October war clearly has been that America's self-chosen role as mediator in the Arab-Israeli question has time and again blunted

What Can Be Done?

its role as adversary and interested party in the oil dispute, and vice versa. It is hard, in any negotiations, to be a mediator and a litigant at the same time.

Syria and Egypt, Israel's chief sources of anxiety, do not themselves directly torment America and the West with respect to oil. Saudi Arabia, which does, plays a more cautious role in the political war. Iran, long an ally of America, has consistently made clear its refusal to support those OPEC Arabs who wish the elimination of Israel; yet it has played a far more ambitious role in hiking OPEC prices than have its Arab neighbors. Even though the parties to these two disputes vary considerably in the intensity of their adversary/non-adversary relations, these two issues of grave contention overlap.

Our intention in writing this book has not been to suggest that in the resolution of one of these overlapping crises the world will find the resolution of the other. The successful extinction of Israel in war, by attrition, or by shameful diplomacy would hardly put an end to the worldwide troubles we have described. In the oil game, no Israel, no America, because without Israel America has no cards to play with the Arab members of OPEC. The elimination of Israel, in fact, could easily confirm the Arab OPEC nations' confidence in their capacities to persist in their enrichment, and would display to a watching world an enfeebled West the more vulnerable to further contrived depredations originating elsewhere.

By the same token, a rampaging Israel, striking at its tormentors and possibly reaching out to "eliminate" the immediate sources of its troubles—Cairo and

Damascus—might thereby bring into play forces which no one could control. In a renewed war, the weaponry will be more swiftly employed, more sophiticated and destructive than before.

But if a war again erupts—as it very well may—when diplomacy collapses, entirely new prospects come into view.

As such a war would very likely bring about renewed embargo, the contradictions between the U.S. role in the Arab-Israeli conflict and the energy crisis then would be greatly heightened. Then, a concern about the enlargement and widening of war would naturally be countervailed by the gigantic, worldwide effects of an oil cut-off. Not wishing to deal simultaneously with both aspects of the situation, the United States conceivably could choose to deal with the one—namely, the Arab-Israeli conflict—effectively, to the deliberate, if temporary, neglect of the other.

The first victims of such American neglect would be America's European allies and Japan. It would be *their* economies which then would chiefly bear the brunt; *they* would face imminent collapse; the folly of past vacilation, of their unwillingness to protect their vital interests collectively, would soon be evident to all. Then they would have to consult their interests; they would have to consider the heavy question of how to defend these vital interests by themselves. The terrible costs of past appeasement would be there for all to see. They would be known to every family of Europe. Adversity confirmed by harsh experience might even bring home to them the necessity of a common position.

What Can Be Done?

The difference between an embargo and the continuing heavy toll of high prices is that an embargo dramatizes victimization by its shock effect. An embargo might encourage solidarity among its victims. Winter cold would take its toll, and everybody would be in the soup together. Energy-deniers would become the enemy. Anger would be directed outward.

With this in mind, the Arab OPEC nations might opt for a half-embargo, in order to break consumer confidence. Consumers thus would be made uncomfortable enough to realize how dependent they are on oil, but not desperate enough to take armed action. But this being the second time around, the consumers might well discover that half an embargo is definitely not better for them than a whole one. A half-embargo is a contradiction in terms. OPEC would find it much easier to manage a total shutdown than a partial cutback. For in the latter situation, enemies as well as friends would somehow get their share—the former, since they would certainly include nations with greater capacity to pay.

A full embargo is not an incremental change; it is a clear and radical restructuring of the rules. Not every oil exporter will be willing to risk everything on a single throw of the dice. Especially if the cause is Israel, Arabs and non-Arabs in OPEC would calculate their interests differently—Arabs and Persians, certainly. If the object is to safeguard a single member, its survival might seem not worth the price to many members of OPEC. Changes of government take place all the time; OPEC cannot interest itself in internal involvements without sapping its strength.

OPEC is kept together by one thing and one thing

only: pushing the price of oil as high as possible and keeping it there. It is in the business of making money by selling, not by preventing people from buying. Production is most likely to be halted when profit is threatened by economic counterpressure—which does not allow OPEC to enjoy its income—or by military action, which would selectively attack certain owners of its income. If armed intervention proved unsuccessful, the embargo would be unnecessary; the exporting countries would continue to dictate their own terms as they do now. But if military measures were successful, the embargo would be unavailable as an option because it requires resources (oil) which would have fallen into other hands. Unless all (or most) of OPEC's exportable oil were in the same hands, an embargo could not endure. The only answer to force is force: but who would be willing and able to use it and who capable of countering it?

The options available are not ones to be plucked from the sky, as Merlin claimed the ability to summon spirits at will to do his bidding. Nor are options unrelated to the concrete circumstances of their use. The happiest circumstance in which the United States soon may find itself would be one in which a certain peace between Arabs and Israel would be attained by diplomacy, coupled with a clear indication that the storm damage occasioned by the oil price was abating. Whether the abatement resulted from collapse of OPEC's prices or from other positive developments would be beside the point. The combination would remove any need to consider the use of force. One would like to think that many in the Arab-OPEC world, as elsewhere, have

What Can Be Done?

given careful thought to the long-range desirability of such a combined outcome.

It is, however, important to distinguish the different states of affairs which the United States might confront, from the best to the worst, from the least to the most risky. Certainly the gravest circumstance would come about from renewed Middle Eastern war coupled with renewed embargo. Then, the two problems would be fused together as one, for everyone to see.

One policy for the United States, then, would be to back-stop Israel, providing means for its survival in combat, and clear military guarantees after a cease-fire. West Europeans (and Japan) would endure, as best they could, the effects of the embargo until such time as it ceased. In effect, American policy would *declare* the two questions to be separate, gambling that the effects of doing so would not occasion the collapse of its allies' economies but rather would inspire them to undertake effective attention to their troubles. The American force thus would sequentially entail military resupply to Israel, followed by cease-fire, followed by U.S. force-emplacements to guarantee whatever territorial arrangements were then established.

Peace can be perilous. The American people may not be willing to support troops in Israel, whose people may also have doubts about the arrangement. Both America and Israel will then have to choose between the dangers of military guarantees and the dangers of war in the Middle East. If the United States is to keep its troops out, it must ward off intervention by the Soviet Union. A stalemate is splendid for the U.S.S.R.—it keeps oil prices high, Israel and America in turmoil, and Arabs dependent on arms. For the same reasons, a stalemate is insupportable for the United States, which must have lower oil prices, and for Israel, which requires peace. Inaction carries a high price.

A far more ambitious response, however, would be to use renewed war as occasion for a major direct involvement, to deal with both matters simultaneously. Such an option clearly has its risks. But it is conceivable that certain circumstances might make it quite plausible. The damage an embargo might do, this time, might have widespread punitive effects which could not be borne. The question then would be whether America's allies would join in collective action. That they have means to do so cannot be doubted.[10] That collective action would do the trick, militarily and psychologically, is likely. The outcome then would amount to a major assertion of the West's common will to survive. But that America's allies would agree is by no means clear. What if they would not?

The political calculus of unilateral American action would then be as harsh as the political consequences of American inaction. The temptation to let things slide would be great. Strong currents of feeling might well arise in the United States to let the world economy collapse, and go it alone. The advice of iso-populism could well be the stark admonition to let the world go to hell. It thus would conjoin with the advice of liberal isolationists in eschewing any force whatsoever.

The gamble then would be whether any American administration could muster enough support to strike swiftly and—from a military standpoint, effectively—to

What Can Be Done?

create by itself a *fait accompli*—an "American option." Surely this would be done only if risks which once appeared ruinous seemed later on worth running, compared to insufferable prospects of passivity. The better things get, the less likely anyone would want war; but the worse they become, the more extreme would be the options considered and agreed to. If the crises did not commingle, no nation would contemplate such radical remedies. The United States, with whatever support it could bring along, might directly invade, and occupy, essential oil regions. In effect, a new oil regime would come into being, promising delivery (at much lower prices) to the world's consumers. In such an action, the question surely would be whether the ensuing likely difficulties—such as sabotage, guerrilla warfare, temporary cessation of deliveries—would outweigh the ensuing likely difficulties of no action at all. That such an equation would be hard to establish, that it would involve guesswork of a most controversial kind, is as obvious as the fact that the equation would sternly demand its own solution.

Like other aggressors, the United States could claim that it is acting not only in its own interest, though that is sure, but for world welfare as well. It could set up an international consortium to sell oil at $6 a barrel, with $4 a barrel going to the exporters and $2 a barrel set up as an immense development fund to be allocated in lump sums through the World Bank, the United Nations Development Fund or any other agency set up by recipient poor countries. Oil and the Middle East would no longer be the same issue. Israeli and Arab would be left to work out a regional solution in a context within which Russia and America provide joint guarantees of borders and/or stopping all shipments of arms into the area. All the poor would be better off than they were before, though some not as rich as they might have been. No rich nations would be worse off, though some not as rich as they were going to be.

(In this regard, the question of Soviet intervention appears again. But in such a circumstance as now is offered, Soviet vital interests would not be threatened. America's would. Restraint by Russia would not threaten its own standard of living or impose widespread unemployment and governmental instability among its allies. In the West all such penalties would be suffered.)

Once in, as we learned in Vietnam, it would not be easy to get out. While the United States might choose sparsely populated areas to occupy—such as Kuwait and Abu Dhabi—it would have to deal with continuing guerrilla attacks and attempts to sabotage shipping. A substantial number of troops would be required. The passage of time would perhaps inspire civilian protests at home and possibly lead to low military morale.

Clearly, the option of unilateral force could be exercised only under the most dire circumstances. In the past, America has been accused of aggressive acts, but these have taken place always in the context of apparent aggression by others beforehand, even in Vietnam. But this action would be open and unequivocal (naked, as some call it). It could make sense, if at all, only if vital interests—political, economic and societal—were dramatic and unmistakable. Of course, it would be very much for the better if they were not.

What Can Be Done?

If oil were substantially reduced in price, or if OPEC loaned the difference between the old (pre-1973) and new prices to consumers so that they could amortize the debt, say, over the next quarter or half century, everything might still turn out all right. Calamity would be avoided and sanity restored. A settlement of local problems could be negotiated. But temporizing with the oil crisis is like living with a time bomb. Were it to explode, the blame would rest not only on those who detonated it but also on those who primed it.

The choice the world really faces now is not only between war and peace but between order and disorder—order which will maintain modes of resolving disputes peaceably, disorder that will dissolve them, and force which would seek to impose new systems of order at risk. If war does not come soon, in an attempt to avoid the grave effects of the oil situation, it might come later, under far more disadvantageous conditions, in response to its long-range impact—to hunger, inflation, unemployment and instability.

Should peace prove impossible, who will desire war? Few people wish to raise this question—and no one wishes to answer it—since reasonable men in a reasonable world would prefer to spend their time on more constructive matters. One can sympathize with Secretary of State Kissinger's insistence that it would be better to find questions governments can answer than to provide answers to questions no one should have to answer.

The Great Détente Disaster

NOTES

1. B. Nagorski, New York *Times*, November 29, 1974.
2. *National Observer*, November 16, 1974.
3. A possible exception to this is the United Kingdom; but the strange feature of the British condition lies in the disparity between a happy future when North Sea resources come into play and the scenario of immediate economic disaster which now unwinds. It is no comfort to a man facing imminent starvation to know that ten years hence a bounteous store of delicious calories will be at his disposal.
4. We often think of Cassandra merely as a prophetess of gloom and doom, contrasting her with Pollyanna. The analogy is not quite accurate: Cassandra, given the gift of prophecy in return for her promised favors, was cursed by Apollo when these favors were withheld; she was condemned to prophesy, on condition that no one would ever believe her.
5. *San Francisco Chronicle*, December 27, 1974, quoting from *Voices for Life: Reflections on the Human Condition*, edited by Dom Moraes (Praeger, 1974).
6. In this respect it ought to be borne in mind that in the isolationist 1930s few of the politically influential isolationist political leaders were also antimilitarist. Despite the Nye munitions investigations, and other manifestations of antimilitarism at that time, the American Congress consistently, and with strong majorities, supported defense appropriations even while opposing administration overtures deemed violative of American neutrality.
7. William Schneider, "Public Opinion: The Beginning of Ideology?" *Foreign Policy*, No. 17 (Winter 1974–75), pp. 88–120.
8. *Ibid.*, p. 109.
9. The only significant breach of this understanding was Soviet action during the Cuban missile crisis in 1962.
10. The ill-fated Suez expedition in 1956 is a memory which still inhibits Europeans' self-confidence. But the failure of this mission was due exclusively to strong opposition by the U.S. as bloc-leader of the Western alliance. Had the tables then been reversed, had America unilaterally acted to seize and reopen the Canal, there can be no doubt as to the finality of the outcome.

AFTERWORD

To WRITE is to think on paper; to publish is to think in public. To think in public is to speak in a place where anyone can listen. Writers cannot select their readers; readers select themselves. We have, however, written this book very much with our fellow Americans in mind, for our principal concern has been to display the full nature of the crisis.

Throughout the postwar period Americans have been accustomed to think about our nation's foreign policies in large and remote abstractions. "High Politics," as Luttwak calls these political, paradigms, has been all about the strategic balance, Communism versus the Free World, overcommitment versus undercommitment, and so forth. We have noted our nation's posture in the world on global maps, and have either accepted or protested it. Vivid as the Vietnam war was to a nation of television viewers, the important questions it gave rise to were nevertheless also abstract and remote. Once American troops were finally withdrawn, few protested the vast killings which followed the "peace."

Our new situation sharply differs from that to which we have been so long accustomed. Daily it affects the most intimate and fundamental aspects of our existence. The people in trouble are no longer "them" but "us."

Now we learn how wrong we were, all along, about the nature of the "primary goods" we need for our wellbeing. These we have long thought to be food, clothing, and shelter. Now we know that these are derivative: their existence and our access to them are contingent upon goods more primary than they. Energy, from whatever source, as well as guaranteed access to that energy, is their precondition; it is the "essential to our essentials." No amount of labor or technology alone can compensate us for their loss. These facts are now known to us in our soaring utility bills, in the prospect of a cold home in wintertime, of less food at higher prices, of fewer visits to distant friends and relatives, and in the penalization of the old and the poor.

These facts also become painfully evident to a government that finds every policy skidding on oil. The president cannot make straightforward prescriptions for increasing employment through tax relief because individual income is simultaneously being decreased by oil price increases. Facile comments on the manageability of the oil deficit founder on a substantial increase in inflation. The government giveth; OPEC taketh away.

For us, oil is surrogate of the very sun itself: *it is the stored rays of the sun.* Until other energy sources take the place of oil, perhaps late in the next decade, we can no more do without it than we could survive in a sunless world.

So it is that the remote and abstract High Politics of the past gives way to a future "Low" Politics, which

Afterword

arises out of the most basic necessities in the lives of very real people, directly and immediately affected by the identical, very real problem. Low Politics will be endowed with few ideals but much self-interest. Some Americans now take the occasion to heap blame upon nearby scapegoats: oil companies, public utilities, food processors, and tax-collecting authorities. But Low Politics might account for our intimate distress more plausibly by identifying the basic source of our troubles. If the sun's rays suddenly diminished, one hardly could blame earthbound clouds; the corner grocer does not "run" the world's agriculture industry; oil companies now do not "own" the almost priceless properties which chiefly supply—or on occasion refuse to supply—the world's oil consumers. What blame there is must be assigned where it belongs.

So far we have found it hard—perhaps because we have been reluctant—to vent outrage where outrage clearly is called for. We have been of two minds in indecision and thus have vacillated about our circumstance.

Low Politics, the politics of iso-populism as we call it, very likely will not concern itself with pious formulations, but very much with mundane profits and losses; in short, with interests rather than ideologies. It will feel less need to apologize for unpleasant action than many politically priggish persons now think, because rhetoric of idealistic isolationists or internationalists alike has no place in our current circumstances; it has become irrelevant. Now, duty follows interest, rather than the other way around.

To some readers of this book in other parts of the

Western world, our analysis may become embarrassing. In truth, America's possible response to the catastrophe deeply affects America's relations with its traditional allies. The economics of it affect them far more directly than they affect the United States. But when one comes right down to it, the hesitant responses of Europe to the new threat to common vital interests is disturbing. The impulse to prevaricate and obfuscate is stronger than in the United States.

Since World War II, neither Western Europe nor Japan has dreamed of being master of its international destiny; they have accepted the shield of American power in preference to the arduous and very risky course of an independent policy. Only in the 1956 Suez crisis did European states embark on major international adventure by themselves. Scarred by this experience, they have since resolutely shied away from any controversial assertion of authority in dealings with other nations.

The solidarity of the Western alliance has, nevertheless been a stabilizing aspect of High Politics. It has provided a wary consensus among Western nations about the need for maintaining an East-West balance. America's European NATO allies often have disagreed upon the strategic means of responding to the problem of Soviet power; but they have never denied its persistent seriousness. But we should notice now that they have drawn a stringent limit around the range of their alliance obligations and of their possible adversary relations. Alert to the security features of East-West relationship in Europe, our NATO allies have had no com-

Afterword

mon security policy with respect to crisis contingencies arising from other sources. Their means have shrunk. Over the years they have nearly liquidated their overseas bases. The British withdrawal from areas East of Suez, long in progress, is now almost complete. The stabilizing role this power once played in the Persian Gulf, for instance, is almost now a matter of history.

Europe, once mistress of a globe, is now almost out of world politics. Her southern flank, the Mediterranean, is one which only the American Sixth Fleet guards, and on which the Russian navy now moves with relative ease. Today the possibility of Europe as a strong, determined and credible guarantor of a future Middle East settlement—a settlement essential to Europe's fate—is never mentioned.

The effect of oil on Europe's broader security problems cannot be denied. With worsened internal economies, Europeans experience internal polarization. Busy juggling petro-dollars, they are less able to manage their own financial affairs now than at any time since the late 1940s. Also they are victims of the neglect of their vital interests. Fearful of reprisal for even a slight grimace at their oil exploiters, they are not unlike old Ben Bolt's girl friend:

> Oh! don't you remember sweet Alice, Ben Bolt,
> Sweet Alice, whose hair was so brown,
> Who wept with delight when you gave her a smile,
> And trembled with fear at your frown?

The states of Europe have vacillated between obsequiousness and opportunism. Now they behave as

Churchill once described Stanley Baldwin's behavior: "Decided only to be undecided . . . adamant for drift, solid for fluidity. . . ."

Europe's actions will be governed by Europe's interests. They certainly should be. But the day on which America disappeared from the Eastern Mediterranean would be far gloomier for Europe than for the United States. It would signify Europe's final vassalage to the Soviet Union and its local dependencies. Our allies might well ask whether they will do better or feel safer if America were to learn the lesson of national self-sufficiency that OPEC's actions and their response are bound to teach.

Where will its "low" interests take America's high diplomacy? "Iso-populism" would react to the turmoil outside the United States by turning inward. It would try to isolate itself against foreign fevers. If not, iso-populists would try to cut out the source of the malignancy so they could quickly revert to domestic cures. Troops would be withdrawn from Korea and NATO. If the Japanese and Europeans were to start to worry, they would be left to defend themselves. An Arab-Israeli war would be left to sort itself out. Should the Soviet Union intervene, and should this intervention appear threatening, American iso-populist leaders would counter the threat with nuclear weapons—having already cut back on regular forces, both to save money for social services and to reduce the risk of international entanglement. Iso-populism would be characterized internationally by periods of quiescence punctuated by fast and furious action. It would be devoted to domestic development, trading when necessary, but enlarging

Afterword

the home market whenever possible. Whether a nineteenth-century policy can prevail in a twentieth-century world may be doubted, but the iso-populist vision of justice at home and, if not exactly peace, at least noninterference, abroad is appealing.

It is just as easy to understand the appeal of such an attitude as it is to demonstrate its unsuitability. International insecurity is bound to increase if the international system is subject to random and unpredictable shocks of great magnitude. "Nuclear war or nothing" is not likely to succeed as a strategic principle. Circumstances are bound to arise in which the United States would rather do something in between. The separation between foreign and domestic affairs, upon which iso-populism depends for its distinctions in theory, is bound to be difficult to maintain in practice. Yet iso-populism can refurbish its image; dressed in different garb, it can now appear as the desirable and defensible "new nationalism."

No folly about isolationism and no fancy words about justice would fit a plainspoken new nationalism. Its vocabulary would be one of narrow, practical interest, the self-interest of all nations and the national interest of the United States. Every nation would be expected to serve itself; no nation, including the United States, would be expected to give or get something for nothing. The transition to the new nationalism would not be abrupt and unexpected. It would be smooth and easily explained; it would flow from words defining concrete national interests and actions designed to secure them.

Would a South Korea in the hands of its northern totalitarian foes be a dagger pointed at Japan and other Southeast Asian countries? The United States would not define Asia's interests for particular nations. They would have to do that for themselves. If a particular Asian interest coincided with an American national interest, the interested parties would try to agree on a sharing of burdens. In no case would the United States alone station troops to defend a region whose member nations lack interest in their own defense, an interest expressed not only in words but in deeds, not only in money but in men and material.

A similar principle and process of interaction might also govern relationships with NATO. The United States would seek to shift defense burdens. Only if its proposals failed would it initially remove part of its forces and then later the remainder. Perhaps, once the disturbing influence of American arms is removed, NATO members would better be able to clarify their own interests. If they believe they face no external threat, they are entitled to put their hypothesis to the test. Allowing everyone to learn from his own mistakes (including experimentation with Marxism in Latin America) would be a basic tenet of the new nationalism.

Oil could be no exception. Every effort would be made to arrange accommodation with oil exporters and alliance with oil importers to drive down the price or obtain long-term loans. If the importers could not agree, or if the exporters did not act, the United States would assume no common interest existed. It would not attack anyone. Instead, it would act on its own behalf to reduce consumption and increase domestic pro-

Afterword

duction of energy substitutes. To the extent energy costs more, America would try to make it up by raising prices of commodities it controlled. The United States accounts for more than two-thirds of world food exports. To the degree its balance of payments was adversely affected, the United States would attempt to increase the price of food exports, caring as little about the problems this caused as others did about its own difficulties. Canada might sharply reduce or cut off its supplies; possibly this could not be helped. But, in the spirit of reciprocity, Canada would be given reason to calculate its interests differently in the future.

Israel would be isolated. It would not be sacrificed for oil because that would only make OPEC more obstreperous. Israel might be used to obtain oil; more likely, it would be expected not to impede a settlement. Israel would not be allowed to involve America in a war it did not wish to fight; neither would it be told to sacrifice itself. Israel would be expected to defend itself in its local context. A Soviet invasion, which no one expects, would be thwarted; so would an Israeli invasion. America would not pay the cost of an inch of Israeli territory beyond its 1967 boundaries. Tit for tat, but no more than that. Nor would Israel be wise to remind America of old debts. "What have you done for me lately?" is the likely reply of the new nationalism.

If this kind of nationalism appears to be something new, that is only because internationalism is as old as the Second World War. Renamed "neointernationalism," internationalism can perhaps offer a promise to learn from its past errors. Otherwise its premises—that the world arena is a better place with America in than out of it and that, for better or worse, America, by virtue of size and capacity, is wedded to the world—are not so much objects of controversy as subjects of boredom. "Yes, no doubt, but what difference does it make?" The attraction of "isopopulism" and the "new nationalism" is that they appear to contain general principles for guiding behavior whereas the old internationalism, tarnished or refurbished, is just a jumble of special cases requiring the exercise of judgment, instance by instance, and we all know where that has got us. The very idea is as tiresome as it is essential. What, other than more of the same, would a neo-internationalism have to offer?

If a lightly armed Japan is preferable to one that is heavily fortified, courage advises acceptance of the actions—keeping troops in Korea and maintenance of America's nuclear shield—conducive to such a comforting condition.

Remaining in NATO is worse than anything except being out of it. NATO is one of those families whose squabbles are better than breakup. It brings together most of the world's democracies, an idea which may seem senseless until one imagines what international life would be like were they permanently parted—alone, afraid, and exceedingly nervous. Nurturing NATO is a nuisance but it is better than becoming an international Nervous Nellie.

NATO is typical of most all alliances: the pursuit of individual interest does not always lead to achievement of collective good. Part of the problem is that the United States is so large a portion of the whole that the

Afterword

other members are tempted to take advantage by shifting the burden to the one party that has shown its willingness to pay. A related problem is that NATO members are differently situated in regard to sources of energy and opportunities for securing supplies; hence their willingness to run risks also differs. The present problem is paradoxical: unless the United States acts in regard to oil, the rest of NATO cannot; when the United States acts, its NATO allies are not needed; they can wait to see how things turn out before committing themselves; minimizing the risk to NATO means maximizing the risk to America. And a neointernationalist America, conscious of the irony, expecting to be denounced (at least until the benefits are distributed), prepares to act out its designated role.

An internationalist foreign policy requires an international economy. The existing economy of oil, with its high prices and encouragement for countercartels in other countries, will, as we have argued, lead to the withering away of world trade. This must be changed. Oil policy, like charity, should begin at home.

The United States should drastically reduce its imports of oil at a rate not less than one-third a year. Simultaneously it should seek to lower oil prices by 50 per cent to around $6 a barrel and to arrange for long-term loans to oil importers from oil exporters to finance the foreign exchange. No one, including America itself, will take American policy seriously unless it shows seriousness through sacrifice. Should these efforts fail to achieve their intended effects, extra economic measures would be undertaken.

Stockpiles of oil can be purchased and reserves

within the United States can be used to lessen the effect of embargo. Negative interest rates on short-term deposits would force oil money into longer-term uses. Restrictions on investment would give oil importers greater control over the use of exporter funds. Recycled oil dollars can be taxed at higher rates. All these devices add up to mitigating the economic impact of oil. They would threaten to wipe out the benefits of a multiple increase in the price, thus threatening the very rationale of OPEC, whose ultimate weapon, as always, would be embargo.

A neointernationalist foreign policy ought in the first instance to aim at heading off confrontation with OPEC. Unity among oil importers would be sought to persuade oil exporters to compromise. "Live and let live" would be the motto: oil exporters retain half a fortune and let importers live on the other half. We fear, however, that such a policy will not succeed because OPEC has too much to lose and its adversaries are insufficiently united on how best to make good their losses. Economic and diplomatic policies must be supplemented by military plans.

What would be the consequences of failing to act before the oil price increase had achieved its effects? The chance that in last-minute desperation a spasm response will be more violent and more likely to spread to confrontation between the nuclear powers should not be minimized. By then, the stakes would be higher for everyone. Nuclear power plants, acquired by oil money, can be converted into nuclear weapons. To the possibility of military miscalculation over oil, therefore, will be added the increased probability of ac-

Afterword

cidental explosion or deliberate provocation leading to a chain reaction. Democracies may by then have fallen, replaced by fascist regimes that worry the Soviet Union, or Communist parties that disquiet the United States. Tension between the great nuclear powers would be heightened by the existence of economically strong but politically weak oil states: the temptation to solve great power problems in one fell swoop, by taking over these oil banks, would be hard to resist. If an "iso-populist" America failed to make up in food what it lost in energy, it might be tempted by the possibility, however illusory, of getting in, grabbing, and getting out. Caught between one extremist who demanded an eleven times increase in oil prices, and another who demanded equal treatment for food, Third World countries would become not merely poverty stricken but perpetually poor. This is not the kind of world in which we would like to live, if we were to be alive to live in it.

As the time of affirmation draws near, it is not surprising that mechanisms of denial become more prominent. Who says a catastrophe is coming? Each day is remarkably like the one before. Change is hardly noticeable, especially for the more affluent opinion leaders, whose jobs are not at stake and whose income absorbs inflation. The professoriat, unlike the proletariat, has tenure. From the same economists who brought us depression and inflation comes a similar message: prosperity is just around the corner. Market forces, they say, will reduce prices as surely as supply will exceed demand. Why market forces that did nothing to increase the "take" price eleven times should suddenly begin to operate is nowhere explained because the logic is inexplicable. The logic of administered prices would work in precisely the opposite direction, for as demand weakened, production would decline and prices would rise so as not to affect total income.

Even if the short term is ugly, the voice of denial insists, the long run will be beautiful. There will be, if not a chicken in every pot, at least a tankful in every garage. In five to eight years it will all be over, the dark storm clouds will pass and brilliant sun will shower an abundance of energy on a radiant world. That the next life should be better than this one, especially for believers, we do not deny. Their messiah indeed may reveal the hidden treasure of the earth in the next decade. What we want to know is how we unbelievers are going to survive until then, and what the necessary measures will be doing to us in the meantime.

But why—the desire for denial keeps making us insist—should the United States be so worried when the nations in the Third World, which suffer the most, and Japan and West Europe, which suffer next, say OPEC is treating them well? Of course, they have to say that. When they are held for ransom, victims quite often say their kidnappers are treating them well.

We, too, have hopes. We hope that as time passes the Western alliance will develop a common view of what is happening to them and a concerted plan of action for changing their circumstances. We hope that a combination of European accommodation and American threat will persuade OPEC to halve its prices and double its loans. We observe that in September 1974, when President Ford and Secretary of State Kissinger pleaded for OPEC to alter its behavior, they were met with con-

Afterword

tempt, but when they asserted the nation's ultimate right to self-defense if it were "strangled" by denial of oil, OPEC became more respectful, saying it would be better to talk things over. We hope so.

For three-quarters of a century now, America has supported a status quo in world politics because it has very much to lose were it overthrown. So it continues. But the status quo is exactly that—it hardly includes the processes which point to its overthrow. As we have suggested before, at some point the Western need for oil could outweigh the fear of using collective force. If the United States clearly were disposed to share oil with its allies in a new crisis, the opportunities for alliance would grow; and a new common readiness to sacrifice would be inspired. But even without the opportunity for common alliance policy, a neointernationalist government would prepare at home for what it might need to do abroad. Other nations have vital interests; is providence so kind as to exempt us from having to care about our own basic needs?

Intervention must be seen as one option, and its results, costs, and benefits weighed against the already great damage resulting from the man-made situation now imposed on us. That damage has been very heavy, continuing, and increasing.

Recently, because of setbacks in our foreign policy, some Americans have concluded that the United States has no vital interests worth defending in the maintenance of an orderly international system. But for a quarter of a century the survival of this system has been largely due to the fact that friends and adversaries alike have known just how serious we had been about main-

The Great Détente Disaster

taining it. Since they are now less certain, that order itself is much less certain. Now, as our willingness to act again with determination might grow, so also the disposition of others to treat us seriously will increase.

That order which we speak of is necessary to our civilization, and in large measure it has been of our own making. We must now decide whether it is worth defending.

We think it is.